Audio Production
Techniques for Video

Audio Production Techniques for Video

David Miles Huber

Focal Press
Boston   London

Focal Press is an imprint of Butterworth–Heinemann.

∞ Recognizing the importance of preserving what has been written, it is the policy of
Butterworth–Heinemann to have the books it publishes printed on acid-free paper, and
we exert our best efforts to that end.

Trademark Acknowledgments

All terms mentioned in this book that are known to be trademarks or service marks are
listed below. In addition, terms suspected of being trademarks or service marks have
been appropriately capitalized. The publisher cannot attest to the accuracy of this
information. Use of this book should not be regarded as affecting the validity of any
trademark or service mark.

Betacam is a trademark of Sony Corporation of America.
dbx is a registered trademark of dbx, Newton, MA, USA, Division of BSR North America,
    Ltd.
Dolby, Dolby A, Dolby B, Dolby C, Dolby SR, and Dolby Tone are registered trademarks of
    Dolby Laboratories Licensing Corporation.
Harmonizer is a trademark of Eventide, Inc.
Kepex and Gain Brain are trademarks of Valley People, Inc.
PCM and GLM are trademarks of Crown International.
Synclavier is a trademark of New England Digital Corp.

International Standard Book Number: 0-240-80148-2
Library of Congress Catalog Card Number: 86-63509

Butterworth-Heinemann
80 Montvale Avenue
Stoneham, MA 02180

10 9 8 7 6 5 4 3 2 1

Printed in the United States of America

Contents

Foreword

For many years, audio was little more than an afterthought to the production of television programming. "Throw a mike out there" was the cliché that seemed to characterize the importance that sound quality played in video production, as far as many were concerned. Another well-known cliché was referred to as the "three-inch speaker theory," which meant that the quality at the end point dictated the effort expended in getting there. Fortunately, this mentality has changed. "The marriage of audio and video" has become the new cliché since the late seventies and early eighties, when music videos, hi-fi VCRs, stereo television broadcasts, and other such enhanced video offspring began leaving their mark on our culture. The early work of engineers and producers to stimulate the courtship resulted in a great deal of valuable and entertaining programming, and these professionals were spurred on by an avalanche of equipment and techniques to assist their efforts.

Any student of audio or video recording must realize the importance that a mutual understanding represents to those in both fields. It is in large part a response to the demands of today's market, as well as the door to a creative playground for innovators and trend setters. To video professionals, this is becoming obvious, as increased demands for audio knowledge are confronting them daily. For audio people, this is indeed a new ballgame. The days when recording studios aspired to produce record albums as their ultimate goal have long since passed. Today, the studios that stay on top are the ones that have mastered the art of blending sounds and images.

This book addresses the importance that audio plays in the world of video production. It is a well-rounded assessment of the equipment, techniques, and technology necessary to work in today's hybridized world of media production. Not only does it deal with the relationship between the two fields, but it backs this up nicely with material on the foundation of each technology. Thus, a terrific education in the history and basics of audio recording, video recording, digital audio, and editing is presented.

David Huber is a scholar who specializes in the cross-pollination between audio and video. His remarkable depth of insight is apparent here in his thorough explanations of the processes involved in recording. By laying groundwork in the separate domains of audio and video, Huber leads the reader logically into a near effortless grasp of the importance they have with respect to one another. The information in this book is written so as to be plainly understandable and is not obscured by tech talk. It is accessible at the level of the practical user and is not off limits to those who do not have a strong technical background.

Audio Production Techniques for Video is a book for video professionals who want to stay in step with the times. It's also a book for audio people who are looking to the future. And it's an essential piece of reading for the student of recording who wants a serious, real-world view of the marketplace and its opportunities.

Keep reading,

David Schwartz
Editor/Publisher of *Mix* magazine

Preface

The intention behind this book is to bridge the gap of knowledge that exists between the professional audio and video communities with respect to the production of stereophonic and/or high-quality audio for video. It is written for both the audio professional who wishes to learn the basics of video production and postproduction and the video professional who wishes to learn the basics of professional audio and multitrack production.

This book introduces the reader to:

- Stages of production and postproduction encountered within the audio and video media
- The audio tape recorder and the video tape recorder
- The basics of synchronization, time code, the edit decision list, and edit controllers in video and audio postproduction
- Fundamentals and equipment of audio production for video, microphones and miking techniques, and stereo miking techniques
- Fundamentals and equipment of audio postproduction for video, the various stages encountered in audio-for-video postproduction, the sound effects library, electronic music production, signal-processing and effects devices, and the audio production console
- Electronic editing techniques, the video edit, and digital audio editing
- Prevention and troubleshooting of time code problems
- Definitions and standards of SMPTE, VITC, and Pilotone

This book is primarily written for the professional user within the audio or video marketplace who is seeking a better understanding of how to handle the demands of audio production for video and who wishes to increase business potential through the development of new marketable skills. It is also aimed at the student of audio or video production who is faced with an evolving industry and is in need of knowledge as to how these two technologies relate to one another in a practical sense.

DAVID MILES HUBER

This book is dedicated to the diversity of the human spirit and to its peaceful celebration.

This book is also dedicated to a very good friend of mine, my father, Oliver Wendel Huber.

Acknowledgments

I would like to thank the following companies and individuals for their gladly given assistance. They made writing this book much easier and even fun.

Steve Hill, Otari Corporation; Bruce Borgerson, Studer/Revox of America, Inc.; Dave Talbot, AKG Acoustics, Inc.; Ferdinand Boyce, Northshore Marketing, Inc.; Earl Fleehart, Fleehart & Sullivan, Inc.; Larry Boden, JVC Company of America; Duanne Barr and Dave Harwood, Custom Video Systems; Richard Newman, Sarah Baker, and Doug Dickey, Solid State Logic Ltd.; Richard Sirinsky, CMX Corporation; David Detmers and Tom Gardner, Ampex Corporation; Peter Hammer, Ampex Museum of Magnetic Recording; Joel Levy, Criteria Recording Studios; Lenard Pearlman, Editel–Chicago; David Smith, Editel–NYC; Doug Tomlinson Photography; Curtis Chan, Ron Petty, and Bob Corbino, Sony Corporation; Alan Penchansky, Geltzer & Company, Inc.; Freidrich-Karl Reichardt, Robert Bosch GmbH; Rhonda Kohler, Rupert Neve, Inc.; Dick Dodson, Turner Broadcasting System, Inc.; R. J. "Bob" Reiss, Jr., Sprague Magnetics, Inc.; The Droid Works; Wesley Dooley and Ron Streicher, Audio Engineering Associates; Nigel Branwell, Audio + Design/Calrec, Inc.; Symetrix; Robert E. Griffin, Inc.; New England Digital; Audio Kinetics; Fran Dym, Dym/SR&A, Inc.; Valley People, Inc.; ESL, Inc.; Gotham Audio; Crown International; Jeanne Meade, Eventide; Abekas Video Systems, Inc.; Society of Motion Picture and Television Engineers; Beth Baj, Adams-Smith.

I would like to express special thanks to Jeff Phillips of Otari Corporation, Christen Hardman of CMX Corporation, Russ Berger of The Joiner-Rose Group Inc, Philip and Vivian Williams, and David M. Schwartz of *Mix* magazine for their patience and assistance. Special thanks also go to Jim Rounds for being the most patient developmental editor imaginable, to Greg Michael for helping me get this project and for making me feel right at home, and to Jim Hill for taking a chance on me in the first place.

1 *Introduction*

A change is occurring within today's entertainment media. As in any field, change often takes the form of a shift in professional attitudes or an advance in technology. These kinds of evolutionary forces keep an industry economically and creatively alive. During the past decade, the medium of video has been at the focus of such change and has undergone a monumental period of growth that has had an impact on all the entertainment media.

One of the most recent advances in video technology has been the achievement of a higher degree of quality in sound production and reproduction. This is due to a number of factors, which are at once evolutionary and revolutionary.

The medium of video evolved out of the technologies for both sight and sound reproduction. Since the inception of television, however, greater emphasis has been placed on the visual aspect. Now this bias has slowly begun to change. In the early 1980s, the public was introduced to "simulcast" broadcasts, for which the video signal was carried by way of television and the audio portion went out in stereo from a local FM radio station. Such broadcasts were a novel and entertaining experience for viewers. The music channels, which were distributed by means of video cable networks, soon followed suit by broadcasting simultaneously in stereo, on a continuous basis, over prescribed FM cable frequencies. This bonding of higher quality audio with video was strengthened at the consumer level by the advent of stereo video cassette, video disc, and laser disc formats.

Recently, the Federal Communications Commission and the National Association of Broadcasters have agreed on a standard means of stereo television transmission. This system of *multichannel television sound*, known as *MTS*, functions in much the same way as does stereo FM radio broadcasting, but the multiplexed set of frequencies is transmitted via standard television subcarrier wavelengths.

One might ask what all the fuss is about: isn't this new stereo gimmick just the same old television with an extra channel added? At first glance, it may seem to be just that, but digging a little deeper will reveal that it is a great deal more. To audio recording professionals, MTS broadcasting represents a final strengthening of relations with the video industry, which have long been in the formation stages. This strengthening has already given a major boost to the audio industry in the form of more production and postproduction care, more hours spent, and thus increased budgets. To the video industry, MTS is no less than the beginning of a new era, an era that has also been marked by the advent of the stereo television set, the audio/video component system (Fig. 1-1), and an enlightened, quality-conscious buying public. The video industry is already starting to feel the effects of this birth of a whole new consumer market, and the extent of its impact is as yet uncalculated.

Fig. 1-1. An audio/video component system, the JVC AV-700. (*Courtesy of JVC Company of America.*)

For many decades, the media of film and professional audio have had a more than casual working relationship due to the necessity of sound mixing for motion pictures. However, it has only been within the past decade that audio production techniques have begun to play an important role in video production. During the 1980s, the professional video production house and the music recording studio formed an alliance in order to create the highly popular music video. Special effects feature films having high-quality, stereophonic soundtracks have been another driving force in the creation of the newly emerging field of *audio production for video.*

Coming up with a concise definition of audio production for video is no simple matter, nor is the newly emerging area fully understood by all members of the professional audio community. The reason for this lack of definition or clarity is that two distinct industries are overlapping, and each of them has its own inner workings and unique set of production techniques.

The methods by which an audio professional can approach sound designing for the modern video market can vary as widely as the diversity of working environments that will be encountered. In recent years, the options for audio production for video have been augmented by new techniques offering a greater degree of quality and flexibility. This broadening of scope has been the result of a combination of the techniques of video production with those of the multitrack recording process. The new hybrid production force is therefore better prepared to create high-quality, "total experience" media presentations.

The audio professional who is working on video presentations must be knowledgeable about the major production phases encountered within both industries. In order to learn how modern audio techniques are incorporated into the building of today's video soundtracks, it is first necessary to understand the basic production stages that are used to create any video presentation.

Stages of Video Production

The creation of a video product may be broken down into three basic stages: *preparation, production,* and *postproduction.*

Preparation

It is beyond the scope of this book to cover the preparation stage in any detail; however, it may well be the most important phase of any video endeavor. Proper preparation, including project organization, equipment organization, and preplanning, is by far the best insurance for achieving the highest possible return for the amount of time and money expended on production.

Production

The production phase of any video project consists of the process of transferring live or electronically generated material directly to the videotape medium. This process may take a variety of forms; each of which makes specific technical demands on the video professional. The most commonly encountered types of production environments are the *live video broadcast*, the *video stage shoot*, and the *on-location shoot*.

In the case of the live video broadcast, a production crew consisting of both technical and creative personnel work in an "on-the-air" status to produce a live telecast. The pressures associated with this form of production are often very high, since all decisions must be made on the spot and are final. As a result, live video broadcasts most often originate from within the highly controlled environment of a television studio, where video and audio control rooms provide the greatest possible degree of reliability.

Video production may also take the form of a video stage shoot. Here the subject matter is recorded directly onto videotape in a controlled environment designed specifically for video production. The usual setup for a stage shoot is a multiple-camera (multicam) one (Fig. 1-2), with each camera connected to its own dedicated video tape recorder, or VTR. This allows the program or event to be recorded on multiple reels of videotape, which constitute the raw footage that will be properly edited together at a later time in the postproduction phase.

Where authenticity is called for in the staging of a video production, it is often advantageous to shoot the required scenes on location. This method of production, which is also known as *remote production* or *electronic field production (EFP)*, requires that all of the necessary equipment be transported to the location site.

A wide range of video production techniques may be employed in an on-location shoot. In such an environment, it is common to record in either single-camera or multi-camera mode without the use of a video control room. As in a video stage shoot, each camera used on location may be set up to record directly on its own dedicated VTR, allowing for decision making to occur in the editing phase. There are,

Fig. 1-2. A video sound stage with a multicam setup. (*Courtesy of Editel Group Chicago.*)

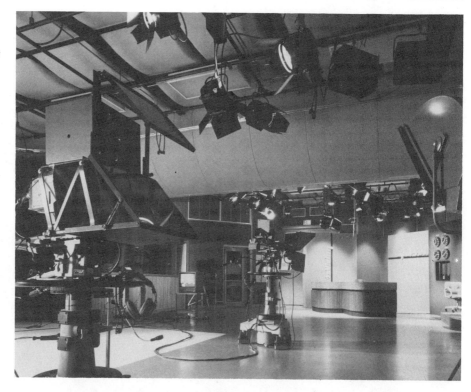

however, some situations in which a remote video control room is necessary when shooting on location. Coverage of a rock concert, for example, will probably be most visually effective if shots are set up and coordinated by a video director. In such a case, a remote broadcasting van (Fig. 1-3) and production team may be sent out to record the video and/or audio portions of the event.

Postproduction

Once the raw production footage has been recorded onto videotape, the final stage of the video project, known as postproduction (also commonly referred to as *post*), begins. The raw footage is often recorded out of sequence, so various camera shots, or scenes, exist on separate reels of tape. These separate shots must be pieced together, in a process known as the *video edit*, in order to create a cohesive and continuous program.

During video editing, a computerized edit controller is used to perform a specified sequence of events (as laid down by the edit list). One or more electronic edits are executed precisely by switching between

Fig. 1-3. Shooting video on location.

(A) A remote broadcasting van. (Courtesy of Turner Broadcasting System.)

(B) A remote control room. (Courtesy of Turner Broadcasting System.)

(C) A mobile audio production unit, the SSL 6000E console. (Courtesy of Turner Broadcasting System.)

the source material from two or more VTRs. The process continues until all of the necessary edits have been performed, at which point the video portion of the project has been transformed into its final version, referred to as the *edited master (EM)*.

Audio Production for Video

The involvement of modern audio techniques in video production may occur during any or all of three stages: the live video shoot, the video edit, and audio postproduction. All of these stages will not necessarily take place in the production of audio for a given video program. The extent of audio production involvement will range from the sound supplied at a live shoot (as for a program shot as a single-camera narrative) to recording for a larger, more involved production, which will include all three stages of development. Many conceivable degrees of complexity are possible, depending on the scope and budget of the project.

The Live Video Shoot

The first step in the audio-for-video process for the majority of projects is the live video shoot. It is here that on-camera dialogue and background sound are recorded directly onto the original videotape *field master* or by a synchronized audio tape recorder (ATR). At this stage, it is extremely important to build quality audio tracks on the original field masters. The goal for the recording of on-camera dialogue or other sound presence is very much like that for the process of multitrack recording—to capture the desired signals on tape as cleanly and clearly as possible.

The Video Edit and Audio Postproduction

The sound recorded during the initial video shoot plays a key role in determining the quality of the final soundtrack, and the building of the audio track during the video edit is of equal importance in assuring that the recorded sound is transferred from the set of original field masters to the final edited master with its quality unimpaired.

System interconnections and controls differ for video and audio, so, during this edit transfer phase, the VTRs and controllers that are utilized must allow for video and audio to be recorded independently. The audio portion may be laid down either before or after the video portion has been transferred; in either case, control is maintained by the edit controller as it is during the video edit.

Since only two tracks are specifically dedicated to audio program material on tapes for the professional Type-C VTR, track building during the editing of the EM is usually done in a monaural or two-track fashion. For larger productions requiring more complex audio, a method of overcoming this limitation to only two tracks is the synchronous transfer of the audio portion of the video field masters onto separate tracks of a multitrack ATR (Fig. 1-4). This practice allows for later changes in level, equalization, and stereo placement to be made, during the process of *mixdown*. It circumvents the generation loss incurred during the audio edit of the EM and makes possible a further postproduction process known as *sweetening*.

Fig. 1-4. The Otari MTR-90II, a multitrack audio tape recorder, within a video edit suite. (*Courtesy of Otari Corporation and AME, Inc., Burbank.*)

Since the introduction of the laugh track in the early 1950s, the use of sweetening as a device for heightening the impact of broadcast visual scenes on the viewers has become quite common. This process may be defined as the addition of recorded material to the basic soundtrack recorded on camera in order to enhance the overall effect of a scene.

Each video project has its own character and set of challenges, so the postproduction process of sweetening will take various forms. The recording or replacement of dialogue, narration, background or foreground music, and/or sound effects adds up to a wide range of techniques and source material that the video or audio producer may draw on in order to create a more effective final product.

The adaptation of the multitrack ATR to high-quality, creative sound recording for the professional video market allowed for a significant increase in flexibility and creativity, as is most evident in today's video productions. The multitrack ATR and its associated modern recording techniques, when used in conjunction with video production and postproduction processes, free up the audio-for-video sound mixer to pay more attention to decisions regarding balance, texture, stereo image, effects, and special effects. Additionally, specific moods may be invoked by the careful placement of a musical or sound effects passage, creating scenes that draw fully on the emotions of the viewers.

2 *The Audio Tape Recorder and the Video Tape Recorder*

In current production practices, the professional audio and video industries rely on magnetic media, generally in the form of magnetic tape, as the major means of storing program material. It is the function of any recording device, used for either audio or video production, to act as a memory device for the storage of massive quantities of information and, when required, to reproduce this information so that it is as close as possible to the original signal. Thus, in theory, a recording device may be looked on as a memory bank with an unlimited capacity for the perfectly accurate storage of separate groups of information within a constant time relationship.

As of this writing, magnetic tape is the most widely accepted means of approaching this theoretical ideal, although in the field of audio postproduction the use of computer-based storage is rapidly developing as the most efficient and cost-effective means for the storage and retrieval of audio information.

In discussing audio production for video, it is necessary to look in detail at the two memory devices most often encountered in everyday production—the *audio tape recorder* and the *video tape recorder* (generally referred to by the abbreviations *ATR* and *VTR*). Any comparison of these two recording devices gives the initial impression of a vast difference in operation and function, but a closer examination uncovers a functional similarity between the two seemingly unrelated machines. Simply stated, the purpose of each device is to record a signal onto magnetic tape, through the use of electromagnetic processes.

A recording system must also operate with a sufficient bandwidth and low enough signal-to-noise ratio so as to be able to reproduce a signal with a quality approaching that of the original. The effective *band-width*—the frequency range between the upper and lower cutoff frequencies of a recording system—is about 10 octaves (20 Hz to 20 kHz) for professional ATRs and 18 octaves (30 Hz to 4.5 MHz) for professional VTRs. This difference in effective bandwidth means that the techniques used for placing signals onto the magnetic storage medium and for retrieving them with a high degree of accuracy will differ to some extent.

This chapter looks in some detail at the basic theory and operating functions of the ATR and VTR. These subjects are not dealt with comprehensively, as that is beyond the scope of this book, but insights and operational guidance are offered for both analog and digital systems, and new developments in computer-based audio and video information storage are discussed.

Magnetic Recording Media

The storage of information for both audio and video production is accomplished by using magnetic tape or, more recently, high-density digital computer disks (or diskettes).

Most magnetic storage media in use today are composed of several layers of material, each serving a specific function (Fig. 2-1). Magnetic tape and computer diskettes often have polyvinyl chloride (PVC) as the base material. This durable polymer has great physical strength and is capable of withstanding a great deal of abuse before serious damage or deformation results. Bonded to the base material is a layer of a *magnetic oxide*, which plays the operative role in the recording process. The molecules of such an oxide form regions called *domains* (Fig. 2-2A), which are the smallest known permanent magnets. On an unmagnetized tape, these domains are oriented in a random fashion over the entire surface of the tape. The result of this random magnetization is that north and south magnetic poles effectively cancel one another out at the reproduce head, giving an average output level of zero. However, when a signal is recorded onto tape, the magnetization from the recording head orients the individual domains into specific directions, so their combined magnetism produces an average magnetic flux force at the surface of the tape (Fig. 2-2B). This average magnetic output may be amplified and further processed so as to reproduce a given signal at some later time with a high degree of accuracy.

Fig. 2-1. Structural layers of magnetic tape for audio and video recording: (1) the topcoat, which is used to smooth the surface of video and digital audio magnetic tape; (2) the magnetic oxide; (3) the polyester or PVC base; (4) the antistatic carbon tape backing.

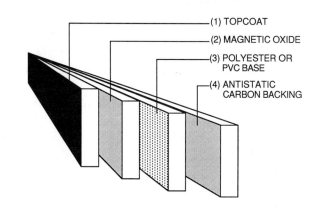

(1) TOPCOAT

(2) MAGNETIC OXIDE

(3) POLYESTER OR PVC BASE

(4) ANTISTATIC CARBON BACKING

There are some significant differences in the composition and thickness of the materials used in magnetic tape for audio and video applications. For example, as will be discussed later in this chapter, the video recording process takes place in a bandwidth of much higher frequency than does the audio recording process. Therefore, the basic tape formulation for video recording must be highly refined, in order to increase the domain particle density and to reduce the instances of dropouts.

Fig. 2-2. Orientation of magnetic domains on unmagnetized and magnetized recording tape.

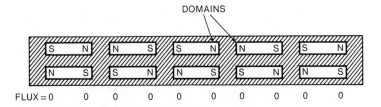

(A) Random orientation on unmagnetized tape results in zero flux output.

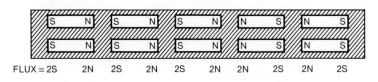

(B) Magnetized domains result in an average flux output at the magnetic head.

The Tape Transport

The process of recording audio signals onto magnetic tape, using a professional reel-to-reel recorder, is directly dependent on the transport's ability to pass tape across a head path at a constant speed and with a uniform tension. This technology is based upon a formula that relates physical lengths of magnetic tape to specific periods of time (Fig. 2-3). During playback, the time spectrum is kept stable by the duplication of the precise speed at which the tape was recorded, thus reproducing the original pitch, rhythm, and duration of each sound.

Fig. 2-3. Magnetic tape recording relies upon the relationship of time to the physical length of recording tape that passes the magnetic tape head.

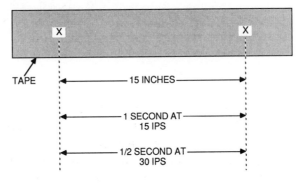

In Fig. 2-4, the transport deck of the Studer A-820 1/4-inch mastering machine is pictured. Note the specific features labeled in the figure: the supply reel, the take-up reel, the capstan, the capstan idler, the tape guides, the tension regulators, the tape shuttle control, and the transport controls and tape timer.

The movement of tape across the head's path at a constant speed and tension is the main function of a tape transport; this function is initiated by depressing the play button. The drive may be disengaged at any time by depressing the stop button, which simultaneously applies the brakes to the left and right reel turntables. The fast forward or rewind button is employed to shuttle the tape rapidly in either direction to locate a specific point. Initiating either of these modes engages the tape lifters, which raise the tape away from the heads. Once the play mode has been engaged, depressing the record button allows the audio signal to be recorded onto the tape. Certain audio recorders require that the record and play buttons be depressed simultaneously in order to enter the record mode; other machines allow for "dropping into" the record mode simply by depressing the record button once the play mode has been initiated.

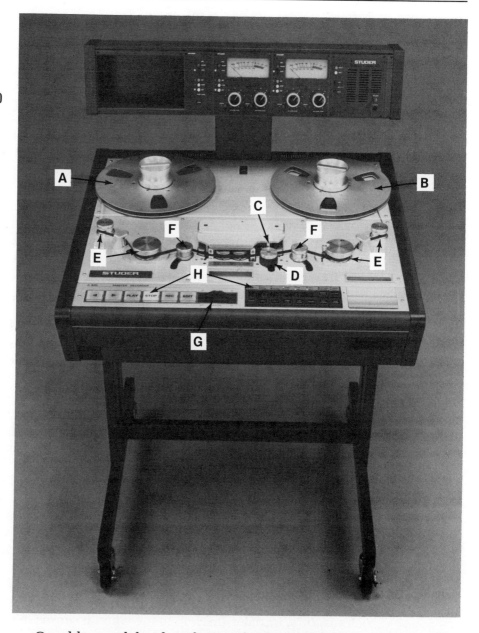

Fig. 2-4. Layout of Studer A-820 transport deck: (A) supply reel, (B) take-up reel, (C) capstan, (D) capstan idler, (E) tape guides, (F) tension regulators, (G) tape shuttle control, (H) transport controls and tape timer. (*Courtesy of Studer/Revox of America, Inc.*)

On older models of professional tape transports, stopping a fast moving tape by pressing the stop button was potentially hazardous to a master tape because of the high degree of inertia possessed by tape moving in the fast wind mode. Sudden stopping resulted many times in a stretched tape or a tape that wound up on the studio floor. To prevent this damage to tapes, a good habit to acquire is taking a simple step known as "rocking" the tape. This is accomplished by engaging the fast wind mode in the opposite direction to the direction of tape travel until the tape has shuttled down to a slower speed, at which time the stop button is pressed.

In recent years, tape transport designs have incorporated *total transport logic (TTL)*, which allows for complete microprocessor monitoring of and control over all transport functions. This innovation has a number of distinct advantages. For example, with TTL it is possible to push the play button while in the fast rewind mode. With an older machine, doing this might literally have stretched a section of the master into magnetic dust particles. But with TTL, the ATR is able to sense that its transport is not at a standstill, determine the direction of tape travel, then either switch the transport into the stop mode or slow the tape to the speed of the play mode, and finally initiate the play mode. Thus, in a limited sense, a tape transport equipped with TTL can be said to be "intelligent." Basically, the development of this "logic-intelligence" is what changed the magnetic tape recorder into the interactive and sophisticated ATR of the 1980s.

Most newer model ATRs are equipped with a *shuttle control*, which allows the tape to be shuttled at various speeds in the forward or reverse direction. This control makes it possible to locate a cue point with the tape lifters off, as well as to wind the tape onto its reel at a slow speed before placing it in long-term storage.

The *edit button*, found on most professional machines, enables two modes of deck operation: stop edit and dump edit. If the edit button is depressed while the transport is in the stop mode, the left and right turntable brakes are released and the tape sensor is bypassed. This allows the tape to be rocked back and forth easily in order to search out edit points. If the edit button is depressed while the transport is in the play mode, the take-up turntable is disengaged, and again the safety switch is bypassed. This dump edit mode enables unwanted sections of a tape to be run off the transport and onto the floor for disposal, while allowing the operator to hear the material that is being disposed of.

A *safety switch*, which is incorporated into all professional ATR transports, initiates the stop mode when it senses the absence of tape along the guide path. Thus the tape is automatically stopped at the end of a reel or upon breakage. Such a switch may be incorporated into the tape tension arm or sensor and may also take the form of a light beam that is interrupted by the presence of tape in its path.

The majority of professional ATRs are equipped with automatic tape counters, which are able to give an accurate readout of elapsed tape time in hours, minutes, and seconds. Many of these counters, which appear as digital displays, are able to double as tape speed indicators, when required in the vari-speed mode.

The *vari-speed mode* allows for variation from the industry standard tape speeds through the use of a continuously variable potentiometer. On many tape transport models, this control provides for speed variations within a ± 20% range.

Capstan Motors

In many tape transport systems available today, the most critical element is the *capstan*. This capstan is the shaft of a motor that is precisely regulated to rotate at a constant rate of speed. There are two commonly found types of capstan motors capable of delivering a high level of speed accuracy: the hysteresis motor and the DC servo motor.

The *hysteresis motor* derives its constancy of speed from the steady frequency of the supply voltage that is used to power it. This standard reference is usually the stable 50-Hz or 60-Hz frequency of the main power lines.

The *DC servo motor* (Fig. 2-5), on the other hand, derives its speed of rotation from the level of direct current voltage supplied to it. During the past decade, this type of motor design has become the standard for tape transports, due to its greater adaptability to systems using motion sensing and TTL integrated circuitry. The DC servo motor essentially owes its stability and versatility to its utilization of *motion-sensing feedback circuitry*. One such feedback mechanism is constructed by mounting a notched tachometer disc directly onto the capstan motor shaft or idler wheel assembly. The rate of rotation is then computed by counting the number of notches on the disc that pass between a light source and an optical sensor per second. Through the use of a resolver to compare the actual clocked rate of rotation with some standard reference, it is possible to obtain a highly accurate and stable capstan speed.

Tape Drive Systems

Currently, three systems for transporting tape across a head path with constant speed and tension are used on professional reel-to-reel audio recorders: open-loop, closed-loop, and zero-loop.

The most common drive system, at least in part because it has been in use the longest, is the *open-loop drive system* (Fig. 2-6). In this system, the capstan and capstan idler work together to move tape along its path at the proper play speed. A small amount of torque is applied at the supply reel motor, in the opposite direction to that in which the tape travels, to provide the required amount of tape tension and good head-to-tape contact. A small forward torque at the take-up motor helps to spool the tape passing through the capstan idler onto the take-up reel.

Fig. 2-5. A DC servo motor with tachometer/re-solver assembly. Slots in the ta-chometer disc allow light to pass through to the sensors for detection of the speed of rotation.

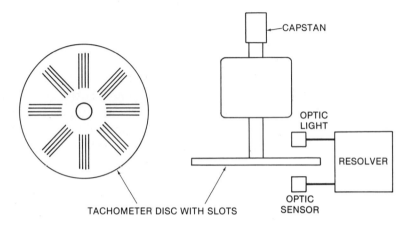

Another drive system used by a few manufacturers, such as 3M, is known as the *closed-loop,* or *differential capstan, drive system* (Fig. 2-7). As the name implies, in a closed-loop system, the tape guide path is isolated from the rest of the transport, with unsupported sections of tape kept to a minimum. This tape isolation reduces many of the distortions associated with open-loop systems. In a closed-loop system,

Fig. 2-6. Open-loop tape drive system.

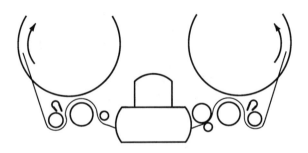

the tape is pulled out of the head block at a faster rate than it is allowed to enter. The entrance and exit of the tape into and out of the head block are controlled either by two separate but synchronized capstans with differing diameters or by one grooved capstan and two pressure rollers with grooves that mate with those in the capstan, providing the same effect as two capstans with differing diameters. The stretching of the tape through the closed-loop system provides the necessary tension required for good tape-to-head contact. Such stretching is possible because of the capacity of tape for being lengthened as much as 5% before permanent deformation takes place. This figure is well within the tolerance of closed-loop drive systems.

Fig. 2-7. Closed-loop tape drive system.

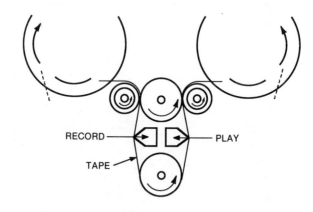

Over the course of this decade, a third system for transporting tape, known as the *zero-loop drive system* (Fig. 2-8), has gained a wide degree of acceptance because it takes full advantage of TTL and motion-sensing feedback circuitry. The zero-loop system does not employ a capstan, as the other systems do. Instead, it makes use of tension and motion sensors on both the supply and take-up sides of the head block to continually monitor tape speed and tension. When variation from the standard reference speed occurs, a resolved, corrective supply voltage is applied directly to the supply or take-up DC servo motor, thus restoring the correct play speed and tension.

Fig. 2-8. Zero-loop tape drive system.

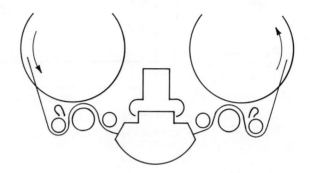

The Magnetic Head

Most audio recording systems employ a magnetic head to perform three specialized tasks: record, reproduce, and erase.

It is the function of the *record head* (Fig. 2-9) to transform the input signal into remanence on magnetic tape for permanent storage. The input current is made to flow through coils of wire that are wrapped around the pole pieces of the head, causing a magnetic force to flow through the pole pieces and directly across the *gap*. Magnetism acts much like electricity in that it flows more easily through certain media than through others. The magnetic counterpart to current is called *flux*, and the resistance to flux is called *reluctance*. The head gap creates a break in and thus a resistance to the magnetic force that has been set up. Since the magnetic oxide on the tape offers a path of lower reluctance than that through the nonmagnetic material within the gap, the flux path goes through the tape from one pole piece to the opposite one. The actual recording of the signal occurs at the trailing edge of the record head, with respect to tape motion, because the magnetic domains on the tape retain the last polarity and intensity of magnetization that they receive before leaving the head gap.

Fig. 2-9. A magnetic audio record head.

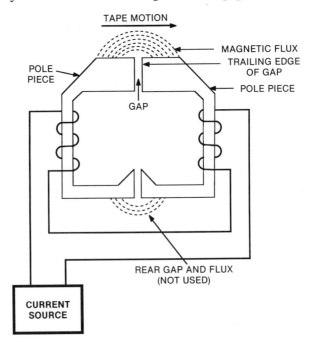

The *reproduce,* or *playback, head* (Fig. 2-10) operates in a similar fashion to the record head, but in inverted order. When a recorded magnetic field passes across the reproduce head gap, a magnetic flux is induced in the pole pieces. This flux induces a current in the pickup coils, and that current may then be amplified or processed.

Fig. 2-10. A magnetic audio reproduce head.

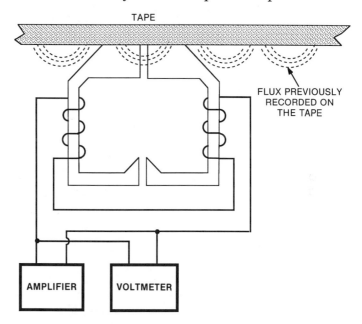

The reproduce head does not operate as a linear device, because its output voltage is proportional not only to the average magnitude of flux presented by the tape but also to the rate of change of this flux. With magnetic tape, this rate of change increases as a direct function of and in direct proportion to the frequency of the recorded signal. Thus, the output voltage of the reproduce head is proportional to:

$$\frac{\Delta\phi}{\Delta t} \qquad \text{(eq. 2-1)}$$

where
$\Delta\phi$ is an average change in the value of the gap flux,
Δt is the time interval required for $\Delta\phi$ to occur.

Given that the output of the reproduce head is directly proportional to the rate of change in the flux, the output of the head will double for each doubling in frequency. This works out to a 6-dB increase in output for each octave (Fig. 2-11).

Fig. 2-11. The effects of increased frequency on the reproduced output at the magnetic head gap.

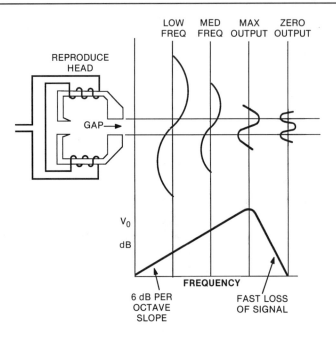

The upper frequency limit of the reproduce head determines the system bandwidth and is in turn determined by the width of the head gap and the speed of the tape. When the wavelength of a recorded signal gets close to twice the width of the head gap, the output signal will begin to decrease until its wavelength is equal to the width of the gap. At this point, the output will be zero, and the upper frequency limit of the system has been reached. The wavelength of a signal recorded on tape is equal to the speed at which the tape travels past the reproduce head divided by the frequency of the signal. Thus, the faster the speed of the tape, the higher will be the upper frequency limit of the magnetic reproduce head. Similarly, the smaller the head gap, the higher the upper frequency limit. Making the head gap smaller has the adverse effect of reducing the playback output signal, thus degrading the signal-to-noise ratio.

The *erase head* serves the function of reducing the average magnetization level of a tape to zero, allowing for rerecording onto tape that contains previously recorded material. When the tape transport is put into the record mode, a high-frequency and high-intensity sinusoidal signal is fed into the erase head, with the result that the tape is alternately saturated in the positive and negative polarity directions. (Tape saturation is attained when all of the magnetic particles at the head gap are completely magnetized and an additional increase in magnetic force will not result in an increase in magnetism retained on the tape.) This alternating saturation serves to destroy any magnetic pattern existing on the tape. As the tape moves away from the erase head, the intensity of the magnetic field decreases and the domains are left in random orientations, with a resulting average magnetization level of zero.

The Analog ATR

(This section deals with the studio production and postproduction ATR. The portable ATR will be further detailed in Chapter 4.) The modern ATR may have any of a number of track configurations, the most common ones being the 2-, 4-, 8-, 16-, and 24-track formats. Each of these configurations is well suited to specific tasks in production and postproduction. For example, 2- and 4-track ATRs are most often used for the initial recording and final mixing stages of audio production and postproduction, and 8-, 16-, and 24-track machines are most often utilized in the creative phases of multitrack postproduction. Some examples of analog ATRs are pictured in Figs. 2-12 through 2-17.

One definition of the word *analog* is "something similar or comparable to another thing in certain respects." When applied to the audio tape recorder, analog refers to the ability to transform an electrical input signal into a corresponding magnetic energy. This energy is then stored in the form of magnetic remanence on tape. The strength of the remanence is in direct relation and proportional to the given input waveform. The analog ATR must have special compensation circuitry, to supply equalization and bias current, in order to operate within its linear limits and with optimal noise and distortion figures.

Equalization

As can be deduced from the previously described characteristics of the reproduce head, the analog recording process is not linear. In order to achieve a flat frequency response curve using magnetic tape, it is necessary to apply *equalization* (abbreviated EQ) within both the record and reproduce electronic circuitry. Equalization is the term used to denote the change in relative amplitude response at differing frequencies.

The 6 dB per octave boost inherent in the response curve of the reproduce head makes it necessary to apply a complementary equalization cut of 6 dB per octave in the reproduce electronic circuitry. This is done in order to produce a resultant flat frequency response curve for the output from the ATR (Fig. 2-18).

Fig. 2-12. Otari MTR-12 1/4-inch 2-channel or 1/2-inch 4-channel mastering recorder.
(*Courtesy of Otari Corporation.*)

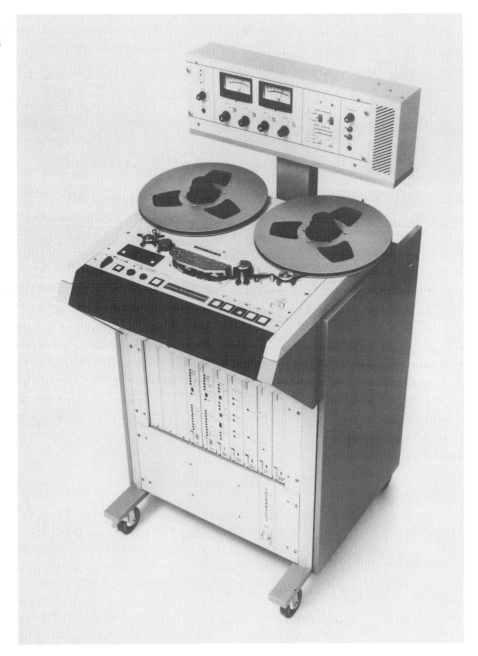

Fig. 2-13. Sony APR-5002 2-channel analog audio recorder/ reproducer. (*Courtesy of Sony Corporation of America.*)

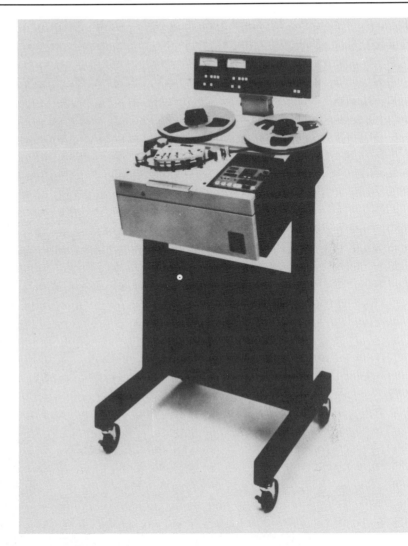

Bias Current

In addition to the nonlinear change in decibel level relative to frequency, there is a discrepancy between the amount of magnetic energy applied to a piece of tape and the amount of magnetism retained by that tape after the initial flux has been removed.

As can be seen in Fig. 2-19A, the magnetization curve of magnetic tape is linear only between points A and B and points C and D. Signals greater than A (in the negative direction) and greater than D (in the positive direction) have reached the saturation level and will be subject to clipping distortion. Any signal falling within the B-C range has too low a flux level to adequately affect the tape's magnetic particles. Thus, some means must be employed to bring the signal at the record head into the linear A-B or C-D range. The means generally used is the application of *bias current*, or *AC bias* (Fig. 2-19B).

Fig. 2-14. Studer A-820 2-channel mastering machine. (*Courtesy of Studer/Revox of America, Inc.*)

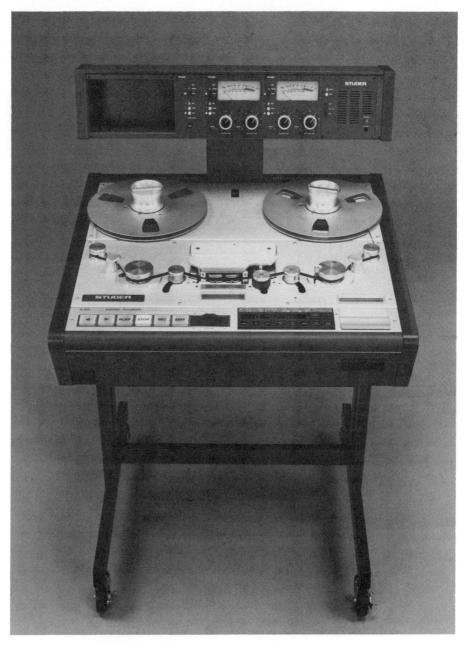

Fig. 2-15. Otari MX-70 1-inch 8- and 16-channel mastering/production recorder. (*Courtesy of Otari Corporation.*)

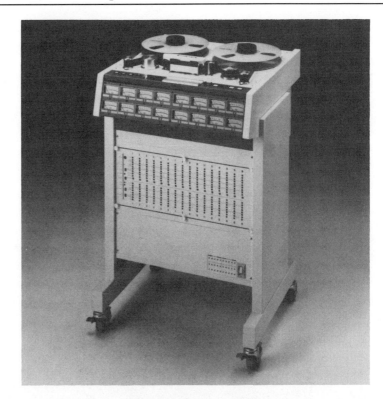

Bias current is applied by mixing the incoming audio signal that is to be recorded with a high-frequency signal (at least twice the highest recorded frequency). This has the effect of modulating the amplitude of the recorded signal to a higher average flux level and serves to move the recorded signal away from the nonlinear zero-crossover range into the linear portions of the magnetization curve. Upon playback, the modulated input signal is reproduced while the bias signal is not, as a result of the reproduce head's inability to play back the high frequencies involved.

Recording Channels

Each recording channel of a modern ATR, no matter what the machine's track configuration, is designed to be electrically identical to the others. The same channel circuitry is simply duplicated the required number of times in order to match the machine's track configuration.

Just as the magnetic tape head is called upon to perform three functions, the electronics of the ATR must be specialized for those same processes—record, reproduce, and erase. In older machines, each recording channel consists of a module in which three separate electronics cards are housed. Each card is designed to perform one of the necessary functions. Newer ATRs, specifically multitrack machines (Fig. 2-20), have an input/output (I/O) module, which is designed so as to incorporate all of the electronics and the adjustment controls of each recording channel on a single printed circuit board. The boards for all

Fig. 2-16. Otari MTR-90-II 8-, 16-, and 24-channel master recorder.
(*Courtesy of Otari Corporation.*)

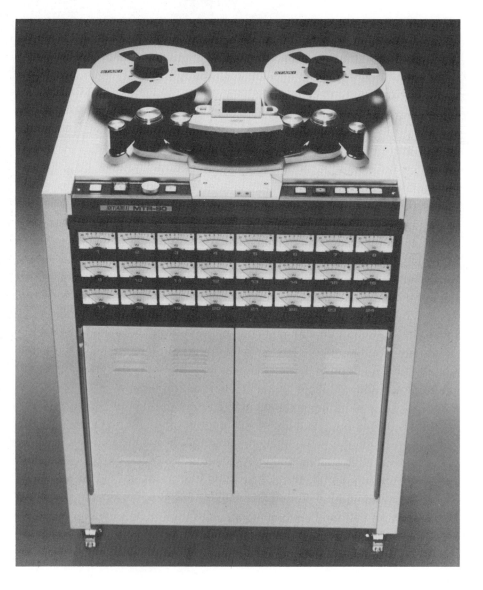

Fig. 2-17. Studer
A-800 24-chan-
nel ATR.
(*Courtesy of
Studer/Revox of
America, Inc.*)

the channels may be housed within a single mainframe that fits inside
the ATR's console housing. Designs using I/O modules allow for channel interchangeability and greater ease of servicing. Modules of either

type provide the necessary level and equalization adjustments for input level, output level, sync level, and equalization trims.

The output signal of a professional ATR may be switched among three working modes: input, reproduce, and sync.

In the *input mode*, the signal produced at the selected channel output is derived from a specific input signal. Thus, with the tape transport in any mode, including stop, it is possible to meter and monitor signals directly from the input of the ATR.

In the *reproduce mode*, the output signal is derived from the reproduce head, with the channel's output and metering reflecting the playback signal. The reproduce mode may be useful in two ways: to play back a previously recorded tape for use in studio production, or to monitor material, off of a tape, while in the record mode. The latter application provides the added advantage of an immediate quality check of the ATR's entire record and reproduce process.

In postproduction with the use of time code, a multitrack ATR may need to be specially equipped with modified electronics. This is due to the fact that control synchronizers often require a reproduce bandwidth of up to 100 kHz in order to read code at higher shuttle speeds. For standardization, the ATR channel with the highest number is fitted with the modified time code electronics.

The *sync mode* is a required feature of the multitrack analog ATR. The sync mode is the answer to the need to record material onto one or more tracks on the multitrack tape while being able to listen simultaneously to the synchronized playback of previously recorded tracks (a process called *overdubbing*). If a recording is made on a multitrack tape while monitoring the previously recorded tracks through the re-

Fig. 2-18. Flat frequency response curve due to complementary equalization within the analog reproduction process.

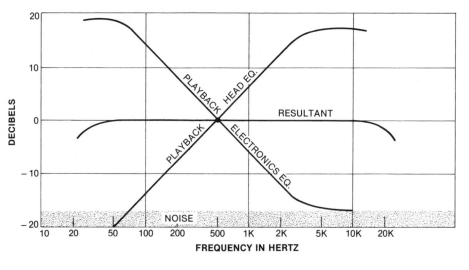

Fig. 2-19. The effects of AC bias on recorded linearity.

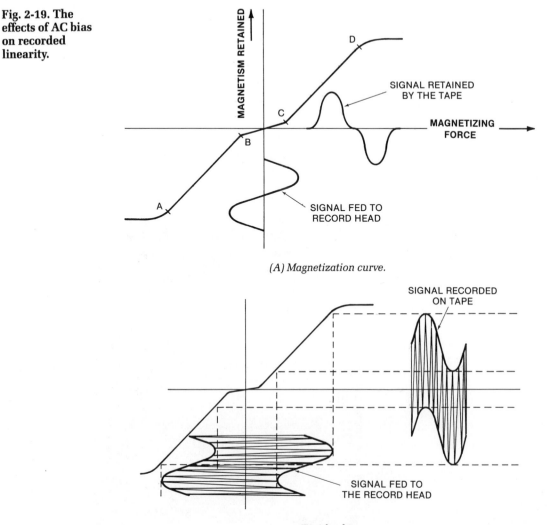

(A) Magnetization curve.

(B) After bias.

produce head, the composite signal, on playback, will be out of sync. In order to prevent such time lags for overdubs, the originally recorded tracks are played in the sync mode. In this mode, the signals to be monitored are reproduced through their respective record head tracks, thereby placing the reproduced and recorded tracks into vertical alignment and maintaining synchronization (Fig. 2-21).

The professional ATR allows for switching each recording channel among the above functions independently. These selection controls may be located on the electronic meter bridge, the front panel deck, or the remote control unit of the ATR.

In addition to the controls over input, reproduce, and sync for each track, the professional ATR has a function selector known as the *record enable switch*. This switch (most often labeled SAFE or READY) serves as a safety device for each recording channel, preventing the accidental erasure of a recorded track.

Fig. 2-20. Studer A800 multitrack recorder with mainframe electronics exposed. (*Courtesy of Studer/Revox of America, Inc.*)

The Remote Control Unit and the Auto-Locator

During the past decade, the remote control unit has evolved from a simple control over deck transport functions to present-day units that contain all of the necessary function and track status controls (Fig. 2-22). These units are most often located near the audio production console, placing the operating controls close at hand for the engineer or operator.

One feature that is built into many of the newer 2- and 4-track ATR transport decks and multitrack remote control units is the *auto-locator*. The auto-locator allows specific cue point locations on a recorded tape to be stored in a memory. Entering a cue point using a keypad or calling one from memory and then pressing the *search button* causes the auto-locator to shuttle the tape to the desired position. Then, the transport will stop, or go into the play mode, or automatically cycle (play-relocate-play) between two cue points. Most auto-locators allow for the storage of multiple reference points, providing for the ready location of the numerous cues encountered in production or postproduction.

Control synchronizers, used in high-level audio postproduction, are also able to operate as interactive remote auto-locators. This form of synchronizer exercises full and simultaneous control over the remote and time code location functions of one or more transports. A specific machine may be manually selected to perform a basic function, or a series of transports may be addressed under time code command. In the latter case, several ATRs may be designated to chase the position of a master tape location or commanded to move tape to a specific time code address. With the event structure that is built into most control synchronizers, it is possible to perform complex record functions with frame accuracy.

Fig. 2-21. The function of the sync mode.

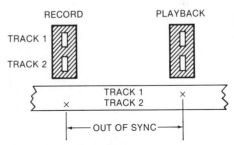

(A) In the reproduce mode, the recorded signal will lag behind the monitored playback signal, creating an out-of-sync condition.

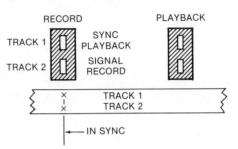

(B) In the sync mode, the record head acts as both record and playback head, thus bringing the audio signals into sync.

Fig. 2-22. The Otari CB-113 remote session controller with optional CB-115 auto-locator.
(*Courtesy of Otari Corporation.*)

Fig. 2-23. Track configurations for basic tape widths used for analog audio recording.

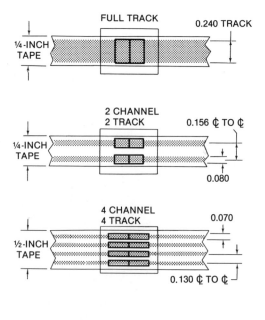

Tape and Head Configurations

Professional analog ATRs are currently available with a number of track and tape width configurations. The most common tape widths are the 1/4-inch, 1/2-inch, 1-inch, and 2-inch (Fig. 2-23).

The optimal tape-to-head performance value for an analog ATR is determined by two parameters: the track and head gap width and the tape speed.

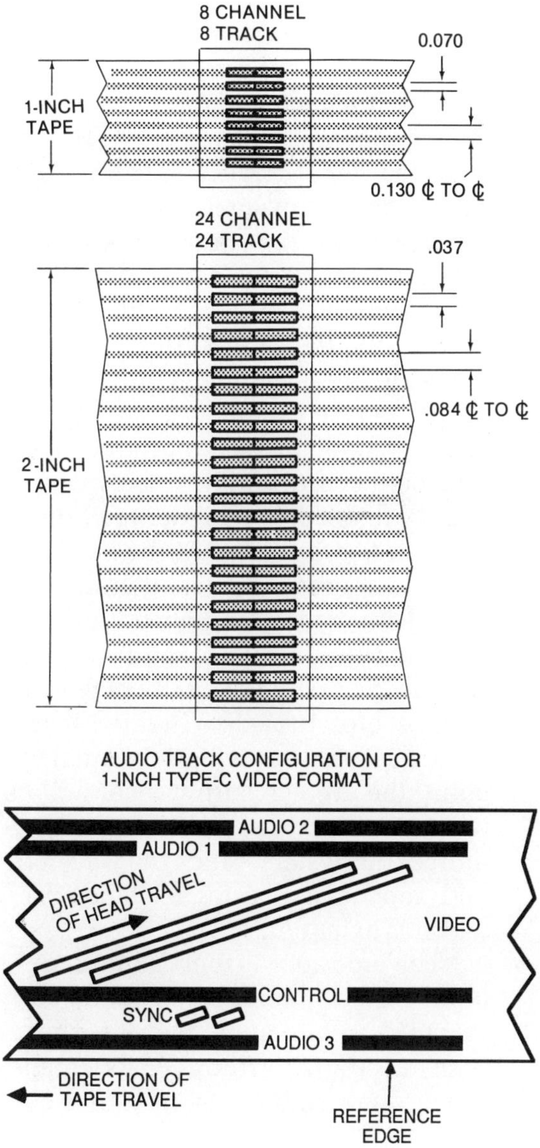

Fig. 2-23. (cont.) Track configurations for basic tape widths used for analog audio recording.

8 CHANNEL
8 TRACK

1-INCH TAPE

0.070

0.130 ¢ TO ¢

24 CHANNEL
24 TRACK

2-INCH TAPE

.037

.084 ¢ TO ¢

AUDIO TRACK CONFIGURATION FOR
1-INCH TYPE-C VIDEO FORMAT

AUDIO 2
AUDIO 1
DIRECTION OF HEAD TRAVEL
VIDEO
CONTROL
SYNC
AUDIO 3

DIRECTION OF TAPE TRAVEL

REFERENCE EDGE

In general, the *track and head gap width* incorporated within professional ATRs is on the order of 0.080 inch for 1/4-inch 2-track ATRs; 0.070 inch for 1/2-inch 4-track, 1-inch 8-track, and 2-inch 16-track ATRs; and 0.037 inch for 2-inch 24-track ATRs. When compared with the 0.021-inch head gap width for cassette tape recorders, these widths are rather large. With a greater recorded track width, an increased amount of magnetism can be retained by the magnetic tape, resulting in a higher output level for the signal and an improved signal-to-noise ratio. The use of a wider track width has the added advantage of making the recorded track less susceptible to signal level dropouts. The *guardband*, a width of unrecorded tape between adjacent tracks, is present in order to prevent channel crosstalk.

The parameter of *tape speed* has a direct bearing on the performance characteristics of the ATR since these are directly related to the recorded signal's level and bandwidth. At higher tape speeds, the average number of magnetic domains that pass by the head gap during a given period of time is greater than it is at low speeds. The average magnetization over a given period of time is therefore also greater, allowing for a higher record or reproduce level before tape saturation is reached. Conversely, the saturation level for slower speeds is reduced, raising the noise floor level. This problem is compounded by the increase of 6 dB per octave inherent in the reproduce head. At slower speeds, the upper frequencies are reproduced at a higher level and reach saturation before the middle or lower frequencies, resulting in high-frequency distortion at higher recording levels.

At faster tape speeds, the bandwidth that can be recorded is effectively increased. This result is also due to the equalization characteristics of the reproduce head. As has been noted, the output of the reproduce head reflects the average value of magnetization at the head gap. As the frequency of the playback signal increases, more and more of the complete cycle will fall within the head gap at any one point in time, until the wavelength is equal to the gap width (Fig. 2-24). At this point, the average output level will be zero. This reduction of output, which is known as *scanning loss*, determines the upper frequency limit of the system. Since the wavelength of a recorded signal increases with tape speed, at higher tape speeds, the upper limit of the bandwidth is extended.

The tape speeds most commonly utilized in audio postproduction are 7-1/2 ips (inches per second), 15 ips, and 30 ips (U.S.), or 19 cm/s, 38 cm/s, and 76 cm/s (metric). These speeds are found on many 1/4-inch and 1/2-inch studio ATRs. A tape speed of 7-1/2 ips represents too much of a compromise of signal quality for use in production work with multitrack ATRs. Over the past decade, 30 ips has gained wide acceptance for analog music production in recording studios. Used in combination with low-noise, high-output tape, this fast tape speed has eliminated the need for extensive noise reduction.

Fig. 2-24. Scanning loss occurs at the point when the recorded wavelength is equal to the width of the magnetic head gap.

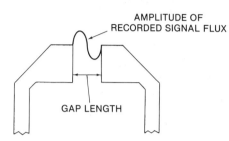

Print-Through
=============

Print-Through

With analog recording, one form of deterioration in tape quality that may take place after recording has been completed is *print-through*. Print-through is the transfer of a recorded signal from one layer of magnetic tape on a reel onto an adjacent layer by magnetic induction, giving rise to an audible false, or ghost, signal on playback. The effects of print-through are greatest when recording levels are very high. The amount of transfer from one tape layer to the next is reduced by about 2 dB for every 1-dB reduction in signal level. The extent of print-through is also dependent on other factors, such as length of time in storage, temperature, and tape thickness (tape with a thicker layer of base material is less likely to have severe print-through problems).

Due to the directional properties of magnetic lines of flux, print-through will have a greater effect on the adjacent outer layer of tape than on the adjacent inner layer. This means that if recorded tape is stored "heads-out" (the tape is wound onto the supply reel), the ghost signal transferred onto the outer layer will be heard as an echo that precedes the original signal. The standard means of storing a recorded tape is therefore in "tails-out" fashion (the tape is wound onto the take-up reel for storage and rewound for production use). With this method of storage, the print-through signal on the outer layer will give an echo that follows the original signal, an effect that is masked by the natural decay of the sound or is perceived by the listener as a reverberation or echo.

Cleanliness

It is extremely important that the magnetic recording heads and moving parts of the ATR transport deck be kept free of dirt buildup and *oxide shed*. Oxide shed occurs when friction causes small particles of the tape's magnetic oxide to accumulate on surface contacts. This accumulation is most critical at the surface of the magnetic recording heads since *spacing loss* can occur as a result of the physical separation between the magnetic tape and the heads due to such oxide accumulation. For example, a signal recorded at 15 ips with an oxide shed buildup of 1 mil (0.001 inch) at the playback head will suffer a playback loss of 55 dB at 15 kHz. This fact illustrates the importance of cleaning the transport tape heads and guides at regular intervals. The accepted method is to clean the ATR using denatured alcohol or an appropriate cleaning solution at least daily and always before routine alignments.

Degaussing

Magnetic tape heads are made from a "magnetically soft" metal alloy. This means that the alloy does not readily retain magnetism but acts as an excellent conductor of flux. However, these heads do retain small amounts of residual magnetism, which can act to partially erase the high-frequency signals recorded on a master tape. For this reason, *degaussing* of a magnetic tape head after 10 hours of operational use is a recommended practice. A magnetic head degausser acts very much like an erase head in that it saturates the magnetic head with a high-level alternating signal that randomizes the residual magnetic flux. Once a head has been degaussed, it is important to move the degausser away from the head at a speed of less than 2 inches per second, taking care not to induce a magnetic flux in the head. Before an ATR is aligned, the magnetic tape heads should always be cleaned and degaussed, so as to obtain accurate readings and to protect the alignment tape.

Electronic Calibration

Since sensitivity, output level, bias requirements, and frequency response may vary considerably from one tape formulation to the next, ATRs have variable-level adjustments for record or playback level, equalization, bias current, and so on. It is extremely important that a standard be adhered to in the adjustment of these level settings, so that a recording made on one ATR will be compatible and identical in response when played back on another ATR. The procedure used to set these controls to a standardized level is called *electronic alignment*, or *electronic calibration*.

Since proper alignment is dependent on the specific tape formulation and set of head blocks being used, it is a good practice to recalibrate all production ATRs at regular intervals. In major recording studios, machine alignments are routinely done the first thing each morning or prior to each recording session, to assure the standardization and reliability of each master tape.

Once the record and reproduce heads are in proper alignment, electronic calibration may be performed. This procedure is carried out with reference to a standard set of equalization curves for each tape speed. Such curves have been established by the National Association of Broadcasters (NAB, in use throughout the United States and much of the world), the Deutsche Industrie Norme (DIN, the standard used throughout Europe), and the Audio Engineering Society (AES, for use at 30 ips).

To anyone new to the process, calibration may seem complicated, but with repetition it becomes second nature and usually takes less than 10 minutes to perform.

The electronic calibration of an ATR is performed in two stages: *reproduce alignment* and *record alignment*.

The primary objective of the calibration process is to reference the levels and equalization of the ATR to prescribed standards. This is accomplished by using a *reproduce alignment tape*. This standard tape, which is available in various tape speeds and track width configurations, contains the following recorded material:

1. The standard reference level is a 700-Hz or 1-kHz signal recorded at a standard flux level (185 nWb/m, standard operating level; 250 nWb/m, elevated level; 320 nWb/m, DIN or European operating level).

2. The azimuth adjustment tone is a signal at 15 kHz for a duration of 30 seconds.

3. The frequency response tones are at 12 kHz, 10 kHz, 7.5 kHz, 5 kHz, 2.5 kHz, 1 kHz, 500 Hz, 100 Hz, 50 Hz, and 30 Hz.

In calibrating the ATR's reproduce electronics, the output switching functions on all channels are put into the reproduce mode. The standard tape is then placed onto the take-up turntable (it is always wise to store this tape "tails-out" so as to achieve an even wind tension) and rewound onto the supply reel. With the controls for level adjustment exposed, switch the ATR into the play mode and adjust for the proper reference level and flat frequency response. The high-frequency reproduce equalization level is best set such that the 10-kHz tone reads at reference level; however, small deviations in response may require special tailoring.

A 7-1/2-ips reproduce alignment tape is similar to a 15 ips one, with the exception that all tones are recorded at 10 dB below normal operating level. This is done in order to allow for the increased saturation of high frequencies at this lower tape speed. The last signal on a 7-1/2-ips alignment tape is a 700-Hz or 1-kHz signal recorded at normal operating level for proper level adjustment.

Most reproduce alignment tapes are originally recorded full-track for all track formats; that is, the recorded signal is laid down across the entire width of the tape. On an ATR with two or more tracks, the reproduction of the low-frequency equalization response cannot be properly set. This is due to an effect called *fringing*. Fringing occurs when a tape of one configuration is played back on a machine with a gap that is narrower than that used to record the signal onto the tape. With this effect, reproduction of signals having a longer wavelength (below 500 Hz) picks up not only the magnetic flux present at the head gap but also extraneous amounts of flux from above and below the head gap width, adding to the signal output at lower frequencies. It is for this reason that these lower frequencies on a test tape exhibit a bass boost, ranging upward to 3 dB or more at 30 Hz. To avoid the inaccuracies due to fringing, adjustments to the low-frequency equalization settings are postponed until after the record alignment has been done.

This process may now be precisely duplicated with the outputs of the ATR switched into the sync mode. By adjusting the appropriate sync potentiometers, all level and equalization controls may also be aligned.

After all of the reproduce alignment adjustments have been made, the alignment tape is rewound and then played or slow-wound onto the take-up reel for storage in the "tails-out" fashion.

Setup for the record alignment stage consists of placing a fresh reel of unrecorded tape onto the machine. This tape should ideally be the reel that is to be used in the actual recording session, but if not, it should at least have the same formulation.

The first setting that is made in the routine adjustment of an ATR's record electronics is that of the *bias adjustment control*. This important adjustment determines the amount of AC bias signal that is mixed with the recorded signal at the record head and has a direct relationship with the noise-to-distortion ratio of the machine. Too little bias signal results in a raised noise floor and a degraded frequency response. Too much bias signal results in increased distortion. Thus, bias provides a compromise or balance between noise and distortion, with the optimum being at some intermediate point.

To set the bias control, once the proper tape is threaded onto the machine, a signal of 1 kHz at 15 or 30 ips or one of 500 Hz at 7-1/2 ips is fed into all inputs of the ATR at operating level (+ 4 dBm/0 VU). With the track output selectors in the reproduce mode, the tape transport is switched to record, and an appropriate adjustment tool is used to locate the individual track controls marked *bias* or *bias adjust*. The ATR's output signal level may then be monitored by means of the VU meter bridge. The bias trim pot is turned counterclockwise until the signal drops. At this point, the trim pot is slowly turned in the clockwise direction until the signal rises to its peak and begins to fall. This is the focal point of the bias adjust. Clockwise rotation is continued until the VU meter reading drops by 1 dB. The bias level is now optimized for the specific tape head and formulation being used. Each recording channel should be adjusted in succession. If the recorded signal begins to peak off the VU scale, the record input level trim pot should be turned back. *Do not adjust the reproduce trim pot once it is set to the standard operating level.*

The additional 1 dB, known as *overbiasing*, has been determined to be the optimal compromise point. It should be noted, however, that it is common practice in Europe to overbias by 10 dB, using an input frequency of 10 kHz. It is claimed that this method gives a highly accurate bias reading.

After the bias adjust controls have been set, the record level and equalization controls are adjusted next. With the tape rewound to the beginning of the reel, a 1-kHz signal at zero on the VU scale is sent from the console to all inputs of the ATR. With all inputs still switched into the reproduce mode, the tape transport is put into the record mode and the *record level* of all channels is adjusted to read zero on the respective VU meters. The ATR is switched into the stop mode and, with all record channels switched to read the input signal, the *monitor level* is adjusted to also read zero on the VU meter for each channel. With these adjustments made, the level between the reproduce and input switching modes is precisely matched.

To adjust the *record equalization controls*, the ATR is placed into the record mode, while being monitored in the reproduce mode, and tone signals of 100 Hz, 1 kHz, and 10 kHz are sent to the inputs at operating level. The individual high- and low-equalization trim pots (marked HI EQ and LO EQ) are then adjusted for the flattest possible frequency response curve.

The Digital ATR

In recent years, the *digital audio tape recorder* has developed from an unstable and expensive recording system to a highly reliable one with unsurpassed technical specifications and an increasingly affordable price. In short, there is little doubt that the digital ATR will eventually become the standard for both music recording and audio postproduction for video.

As described earlier, an analog ATR operates by recording the audio waveform onto magnetic tape in the form of remanent magnetization; the signal can then be reproduced at a later time in roughly its original form. This method of direct transduction means that the nonlinearities inherent in the magnetic recording process, such as modulation noise, wow and flutter, and irregularities in frequency response will also be reproduced, degrading the overall reproduced signal.

The digital ATR operates in an entirely different fashion. The signal information available at the input of the digital ATR is converted into a series of numerical streams of information, which are then stored in a memory device. Upon reproduction, the numerical information is pulled from memory in the precise order in which it was recorded and is converted back into an analogous representation of the original input signal. Since the recorded information is stored numerically, it is not directly affected by the nonlinearities of the recording medium itself. This is due to the fact that irregularities in the recording medium may be compensated for electronically, by applying techniques for time base and error correction. Currently, the usual medium for recording digital information is longitudinal magnetic tape, although the use of computer-style disk storage is becoming more technologically and economically feasible.

Presently, the most widely accepted method of recording information digitally onto magnetic tape is through a form of encoding known as *pulse code modulation*, or *PCM*. The basic theory behind the PCM digital recording system is relatively simple. All basic signal waveforms, whether audio or video in nature, may be broken down into two components: time and amplitude (Fig. 2-25). With PCM, the time component is broken down into a continuous stream of short, precisely measured segments, whose length is crystal-locked to a specific clocking frequency known as the *sampling rate*. This standard sampling rate, which may be set between 44.1 kHz and 50 kHz, serves to break the incoming waveform into a series of discrete steps that are so minute that each one may be represented by a specific voltage level at one instant in time. It is the function of the PCM system to determine the signal level of each specific *sample* and to assign to each one a specific set of binary numbers (called a *word*). This binary word consists of a series of *bits*. A bit may be either a one (represented as a pulse) or a zero (represented as the absence of a pulse). A number of these binary bits are joined together into a word that constitutes a numeric representation of a recorded waveform's signal level at one specific point in time.

Fig. 2-25. Graph of waveform amplitude vs. time.

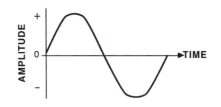

The encoding of the signal information is done by an *analog-to-digital*, or *A/D, converter*. Through successive sampling of the amplitude of the incoming waveform at precisely timed intervals, the A/D converter records a highly accurate representation of the original signal (since the component of time is precisely controlled, the only system variable is amplitude). The digital equivalents, which correspond precisely to the input signal waveform, may then be stored in the form of a stream of binary words on any medium that is capable of storing digital pulses, such as magnetic tape, Winchester or floppy disk, compact laser disc, or integrated circuit memory. Once stored on the chosen medium, these pulses may be reproduced by reading the stream of words at precisely the same sampling rate at which they were recorded. During playback, each digital word is read and reconstructed into its original analogous voltage level by a *digital-to-analog*, or *D/A, converter*.

In digital recording, it is the digital pulses containing the encoded information that affect the final playback quality and not the recording medium itself. This is a major departure from analog recording in that the waveshape of the recorded pulse and the nonlinearities induced in the system by transport irregularities are no longer of consequence. As long as a recorded pulse (a 0 or 1) can be detected accurately by the digital-to-analog circuit, the signal will be reproduced accurately.

Since a digital processor operates by sensing the voltage level at each sample pulse, the PCM system functions by recording and reproducing these levels in discrete steps at each sampling point. Thus, a digitally recorded signal is recorded and reproduced as a series of very closely spaced level steps and not as a continuous waveform (Fig. 2-26A). To overcome this effect in playback, a filter is introduced after the D/A converter; this filter acts to smooth out these discrete step voltages and to restore the signal to its analogous equivalent (Fig. 2-26B).

Fig. 2-26. Sine-wave output of digital processor.

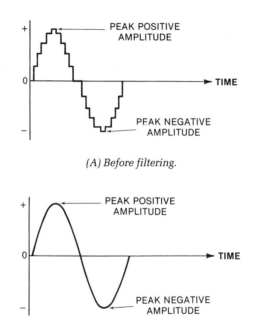

(A) Before filtering.

(B) After filtering.

Performance Specifications

In digital recording, the quality of the recorded signal is dependent on the accuracies of the digital processing equipment rather than on the quality of the stored signal itself. This makes the performance specifications for a digital ATR (Chart 2-1) quite different from those of its analog counterpart.

The signal-to-noise ratio of a PCM system is determined by the number of binary bits that are included within a digital word and by the noise figures for the A/D and D/A converters present in the system. Digital PCM technology is currently based on a 16-bit word structure.

Given a properly designed system, the signal-to-noise ratio (*S/N*) for a signal coded with *n* bits is given by:

$$S/N = 6n + 1.8 \text{ (dB)}$$

(eq. 2-2)

For a 16-bit system, this yields a noise figure of 97.8 dB, which exceeds that of the conventional analog ATR by more than 30 dB.

Chart 2-1. Typical Performance Specifications for a PCM Digital ATR

Dynamic range	Better than 90 dB
Frequency response	10 Hz–20 kHz (±0.5 dB)
Total distortion	0.02% (at peak level)
Crosstalk	−85 dB (1 kHz)
Wow and flutter	Limited only by quartz crystal oscillator
Playback signal-level variation	None
Print-through	None
Residual level after erasure	None

The frequency response of a PCM system is governed by the level detection circuitry within the A/D and D/A converters. In a 16-bit system, the number of possible discrete level steps at any given point in time is 2^{16}, or 65,536. A signal that is recorded and translated into a digital word will be translated back into the same precise voltage level upon playback.

It is important to note that nonuniformities in the level or frequency response of recording tape will not affect the performance of a PCM system, since the tape is simply a storage medium for the recorded binary data. The wow and flutter and sideband modulation noises associated with tape path irregularities are virtually eliminated because the record and playback circuitry is locked in time by the quartz crystal clocking frequency. Upon playback, the digital data are first written into a *buffer memory* for temporary storage and then read out of the buffer in perfect synchronization with the clocked sampling frequency. This has the effect of reducing the effects of tape path irregularities to a level well below measurability. Print-through and crosstalk are also reduced to nonmeasurable levels. Due to the digital nature of the process, the presence of a detectable signal pulse will result in an accurate reproduction of that signal, and the absence of a digital signal due to extreme dropout errors or a data level that is too low will result in zero output due to the error detection and correction circuitry.

Error Correction

The density of the information recorded with PCM equipment is generally extremely high (about 20,000 bits per inch at 15 ips on a fixed-head deck), so that dust or any foreign matter adhering to the surface of the magnetic recording medium will very readily generate error signals. Error-correction codes are therefore normally added to the signal so that, should an error arise, it will be automatically detected and corrected. If the errors are too numerous to be corrected in this way, error-concealment techniques are brought into play. These measure the error-free signals that immediately precede and follow the sample that is in error and substitute their average value in place of the error sample. The effectiveness of these error-correction and error-concealment functions is critical to the stable and reliable operation of PCM equipment. The three major correction or concealment codes are the *Reed-Soloman code*, which acts to correct errors across the track widths of the tape; the *cyclic redundancy code*, which detects errors along the width of each tape track; and the *cross-interleave code (CIC)*, which reduces dropout errors.

Fixed-Head and Rotating-Head PCM Systems

Digital PCM recording systems may be divided into two categories: those with a fixed head and those with a rotating head.

The Fixed-Head PCM Recorder

The *fixed-head PCM recorder* is a reel-to-reel type of recording device, often closely resembling analog recorders in design and function. Mitsubishi has developed two fixed-head PCM recorders, the X-86 and the X-850, both of which utilize the PD (Pro-Digi) format, adopted by the Mitsubishi Pro Audio Group, AEG Aktiengasellschaft, and the Otari Corporation. The X-86 (Fig. 2-27) is a 2-track PCM mastering machine, whose recording format consists of ten tracks along the width of the tape. This format, shown in Fig. 2-28, is made up of two analog tracks (one for monaural tape cueing and one for time code) and eight PCM tracks (six for audio signal data and two for error correction). A combination of the cyclic redundancy code and the Reed-Soloman Code is applied to ensure that all errors in one or both audio recording channels will be corrected. Tape defects caused by dust or dirt will rarely extend across two track widths, so nonrecoverable error signals will not arise under most conditions. When any error beyond the capacity of the correction system arises, it is detected and dealt with in most PCM machines by muting (completely interrupting) the program material. In the X-86, a cross-fade technique is employed, which averages the signal before and after the error point, smoothing out the transition.

One feature of the X-86, which has become standard on most fixed-head PCM machines, is the facility for performing a manual splice edit

Fig. 2-27. Mitsubishi X-86 digital mastering ATR. (*Courtesy of Mitsubishi Pro Audio Group.*)

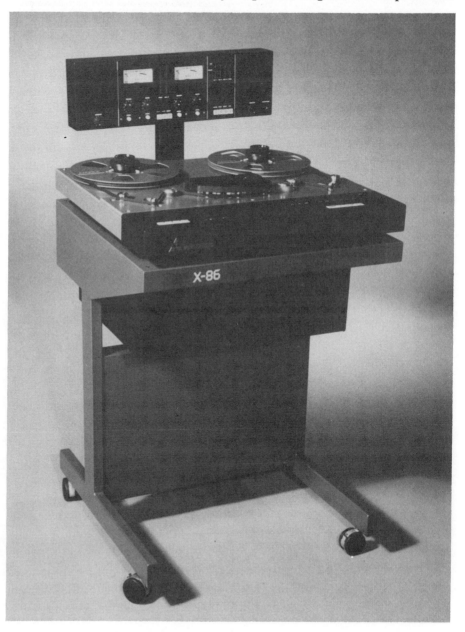

**Fig. 2-28.
Pro-Digi track
format used
within the
Mitsubishi X-86
digital ATR.**
(*Courtesy of
Mitsubishi Pro
Audio Group.*)

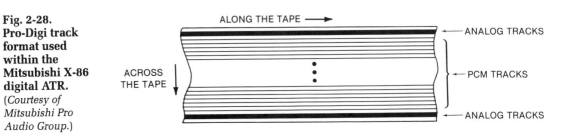

of a tape. The X-86 provides an analog cue track, which allows the manual rocking of a tape back and forth in order to locate an edit point. This cue track may be audited over studio monitors or over an internal amp/speaker combination built directly into the recorder. Once two edit points have been found, a splice vertical to the tape is made by a special editing block to join the two points together for playback.

**Fig. 2-29.
Mitsubishi X-850
32-channel ATR
with remote ses-
sion controller.**
(*Courtesy of
Mitsubishi Pro
Audio Group.*)

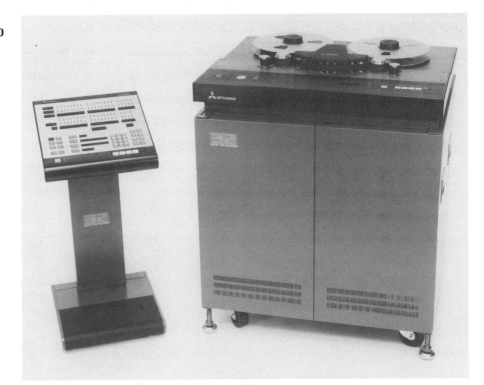

The Mitsubishi X-850 (Fig. 2-29) is a fixed-head, 32-channel PCM recorder, which uses a specially formulated 1-inch tape at an operating speed of 30 ips. All 32 channels are accessible at all times for recording and overdubbing. In addition, three digital tracks and two analog tracks are available for storage of time code or other information. With a high-density data storage capacity greater than 30,000 bits per inch, the X-850 had to have a comprehensive error-correction system to correct the errors that inevitably would arise. Mitsubishi devised what is termed a semiseparate error-correction system. In this system, one audio channel is assigned one digital PCM track on the tape; for error correction, two parity-check tracks are provided for each subgroup of eight channels. This means 40 digital tape tracks are required in order to record 32 audio channels. Each channel also incorporates its own cyclic redundancy code, for detecting errors along the tape's width, allowing for extensive overall error correction of the digital audio signal.

Sony has developed a 2-channel, multitrack digital recorder employing the *digital audio stationary head (DASH) format*, a format jointly announced by Sony, Willi Studer AG, Matsushita Electric In-

Fig. 2-30. Sony PCM-3102 digital ATR. (*Courtesy of Sony Corporation of America, Inc.*)

dustries, and MCI. The Sony PCM-3102 (Fig. 2-30) 2-channel digital ATR is designed to operate at 7-1/2 ips (19.05 cm/s) in a 5-channel I/O configuration: two digital, two analog cue and one analog LTC time code; utilizing 1/4-inch tape, at a sampling rate of 48 kHz (switchable to 44.1 kHz at a transport speed of 17.5 cm/s). The PCM-3202 is similar in design to the PCM-3102 except that it has an operating transport speed of 15 ips (38.1 cm/s) and an additional video control channel. Cross-fade error detection is utilized in both machines to allow for razor blade editing of the master tape, and both models are equipped with a built-in SMPTE/EBU time code generator and chase lock–type synchronizer, providing the intelligence necessary for serial and parallel machine control within synchronization networks.

Sony's 24-track PCM-3324 (Fig. 2-31) utilizes 1/2-inch tape running at a speed of 30 ips, with a 48-kHz sample rate. Across the width of the tape, there are 24 digital tracks sandwiched between two outside analog tracks and an additional control and external data track located

Fig. 2-31. Sony PCM-3324 digital audio multichannel recorder. (*Courtesy Sony Corporation of America, Inc.*)

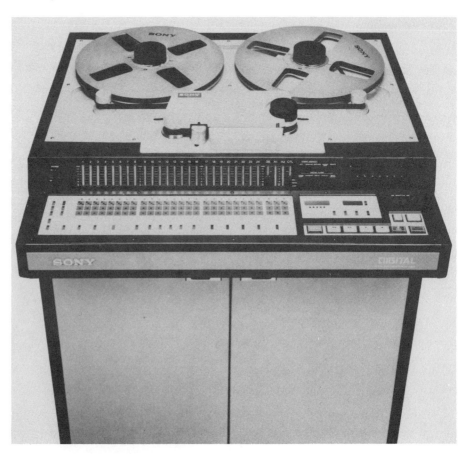

down the center. The error-detection circuitry of the PCM-3324 allows splice editing to be carried out by execution of a 5.20-millisecond cross-fade.

The DASH Format

The following is an outline of the DASH format, which was developed to help ensure standardization between present and future digital ATRs and tape.

1. *Structure*: The format consists of three versions—fast, medium, and slow—and the one that applies is determined by the tape speed of the recorder (Fig. 2-32A). The number of tracks necessary to record a single audio channel is one for the fast version, two for the medium version, and four for the slow version.

2. *Linear packing density*: This applies to all versions. Because of the newly developed HDM-1 modulation code (shown), the minimum wavelength is 50% longer than with conventional codes such as MFM.

   HDM-1 Modulation Code Specifications

   1.51K bits/mm (38.4K bits/inch)

   1.01K flux reverses/mm (25.6K flux reverses/inch)

   λ_{min} = 1.99 mm (78.2 mils) (minimum wavelength to be recorded)

   λ_{max} = 5.96 mm (235 mils) (maximum wavelength to be recorded)

3. *Track density*:

   a. Recording at double the track density can be done using state-of-the-art thin-film heads, while maintaining compatibility with standard track densities (see Fig. 2-32B).

   b. Although it is not included in Fig. 2-32B, the slow version for 1/2-inch tape is a viable choice of application.

4. *Error correction*: The encoders shown in Fig. 2-32C operate based on the cross-interleave code (CIC), with increased interleaving between the even-numbered and odd-numbered words. The CIC can correct errors that correspond to a maximum of three words.

   The interleaving of even and odd words allows splice editing to be done, while protecting against possible damage to the recorded tape. The correctability of burst errors is determined by the encoders and is the same for all versions. Encoding and decoding of the error-correction codes are done independently for each track. If excessive errors appear on one of the recorded tracks (as with a dropout), the correction of the other tracks will not be affected. This feature safeguards a recording during operations under adverse conditions.

5. *Editing*: The functions of cross-fading, punch in and punch out, tape splicing and electronic editing are all feasible with this format, allowing for a natural continuation of the recorded sound.

Fig. 2-32. Aspects of the DASH format.

SAMPLING RATE	TAPE SPEED		
	FAST	**MEDIUM**	**SLOW**
48 kHz	76.20 cm/s (30 ips)	38.10 cm/s (15 ips)	19.05 cm/s (7.5 ips)
44.1 kHz	70.01 cm/s (27.56 ips)	35.00 cm/s (13.78 ips)	17.50 cm/s (6.89 ips)

(A) Tape speed and sampling rate.

TAPE WIDTH	1/4 INCH		1/2 INCH	
TRACK DENSITY	NORMAL	DOUBLE	NORMAL	DOUBLE
DIGITAL TRACKS	8	16	24	48
AUX TRACKS	4	4	4	4
DIGITAL AUDIO CHANNELS — FAST	8	16	24	48
DIGITAL AUDIO CHANNELS — MEDIUM	–	8	–	24
DIGITAL AUDIO CHANNELS — SLOW	2	4	–	–

(B) Track density and channel numbers.

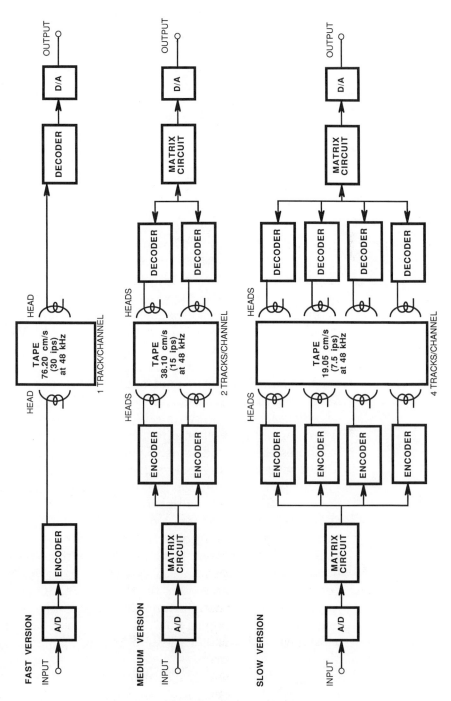

(C) The three tape speed versions of the DASH format (encoders and decoders of all versions are identical).

6. *Future possibilities*:
   a. The advent of the thin-film head can be expected to lead to the development of a number of double-density recorders, bringing increased versatility and economy. The DASH format will make it possible to play normal density tapes produced today on the digital ATRs of tomorrow.
   b. The DASH format is generally applicable to 2-channel through 48-channel recorders, resulting in increased production efficiency and enhanced serviceability. Moreover, with the development of better LSI circuitry, smaller, lighter recorders that are lower in both cost and power consumption can be expected to appear.

The Rotating-Head PCM Recorder

A *rotating-head PCM recorder* is generally made up of two separate components: a PCM A/D and D/A converter and a video cassette recorder (VCR). In 1979, the Electronics Industries Association of Japan established standards designed to ensure the interchangeability of PCM audio recording equipment when used with VTRs (or more specifically VCRs).

In recent years, the price range of PCM/VCR combinations has dropped, allowing even amateur audiophiles to produce high-quality digital recordings at home. Rotating-head PCM/VCR recorders have the advantages of being low in cost, high in quality, and often portable. However, any form of comprehensive editing of tapes made on these machines has to be done electronically, since a manual edit at the VCR would generate errors that would be too great to be corrected for.

One such combination unit is the Audio + Design Pro 701 digital processor (Fig. 2-33), which is a professionalized version of Sony's PCM-701ES 16-bit digital processor. The Pro 701 incorporates a patented coincident time correction (CTC) circuit, which corrects for the 11.3-microsecond delay between tracks in the digital domain, thus assuring a time-aligned output. In addition, the Pro 701 provides functions for switching between the PAL and NTSC video formats, record preemphasis, and copy prohibit. A digital interface also allows both sending to and receiving from Sony's PCM-1610 and PCM-1630 industry-standard digital processors for mastering of compact discs.

Computer-Based ATR Systems

Perhaps the most promising and exciting new technology in the field of audio postproduction for video is the *computer-based ATR system*.

Fig. 2-33.
Audio + Design
Pro 701 profes-
sional PCM
digital audio
processor, front
and rear views.
(*Courtesy of
Audio + Design/
Calrec, Inc.*)

One such system is CompuSonics DSP-2002 (Fig. 2-34). The standard model DSP-2002 comes with a 143-megabyte hard disk system,
which holds up to 18 minutes of mono information or 9 minutes of
stereo information; additional hard disks are available, bringing total
recording times to approximately 11 hours in mono and 5 hours in
stereo. Through the use of data-reduction techniques, these basic figures can be significantly increased while full preservation of the recorded information is maintained. This digital signal processor looks
much like a standard computer with the usual hardware (CPU, CRT,
and keyboard) but is designed to operate as a disk-based audio computer. It is able to transform music, narration, or sound effects into computer data, resulting in an increased ease of manipulation during
production.

Disk-based audio storage systems offer an unprecedented potential
for time savings and flexibility in postproduction, since they are able
to provide *random access storage* for sound effects and music beds.
These units are able to synchronize with any external source (such as
the SMPTE/EBU time code or video sync), making it possible to gain
instant access to any specific sound event under the command of time

code. Thus, at a specific time code address, the computer-based ATR may be programmed to trigger a specific sound event. Since such an event is accessed digitally, it may be called up at any address point and retriggered any number of times. In high-level postproduction work, where speed and accuracy are of the essence, random access storage eliminates the need for multitrack or multiple ATR sound effects bed operation, and in fact provides a higher degree of accuracy and sound quality. Since the recorded information is stored digitally, it is possible to perform edits from widely separated sources with a high degree of accuracy and flexibility. *Effects looping* (the repetition of an effects event) is also definitely improved in execution since the effects of tape shuttling are totally eliminated in the disk-based domain.

Another computer-based ATR system currently on the market is AudioFile, produced by Advanced Music Systems (Fig. 2-35). In its simplest form, AudioFile can capture samples of sound, edit those samples, and store them in a nonvolatile form for recall and playback at any future time. Complete stores of sound effects may be recorded, edited, cataloged, and saved within memory. AudioFile can also assign samples to any of its outputs for multiple synchronous triggering. This triggering can be effected either manually, by auto input, through an events controller, or by using the internal SMPTE time code reader/generator.

AudioFile is also available configured as a multitrack digital recorder. In this form, it offers significant advantages over conventional

Fig. 2-34. CompuSonics DSP-2002 computer-based digital ATR.
(*Courtesy of CompuSonics Corporation.*)

magnetic tape machines in that it is able to slip (advance or retard) any one "track" with respect to any other. What are tracks for a conventional multitrack recorder are digital *files* for the AudioFile. The machine's large storage capacity means that hundreds of files may be stored at once and delivered to any output on cue. Operation can take place independently according to the machine's own internal clock, or it can be locked via the internal time code reader/generator to synchronize with film, a VTR, or another conventional analog/digital ATR.

One major advantage of using a device such as AudioFile is that the "elastic band" effect resulting from having mechanical tape transports locked by a synchronizer is completely eliminated. When playing back a video loop, the audio will be synchronized precisely with the picture virtually as soon as the video image settles in. In addition to the fundamental advantages already noted, it is possible to see advantages in the added digital audio tracks that are provided "invisibly" by synchronously locking the computer-based ATR to a conventional analog or digital ATR. Alternatively, a device such as AudioFile can eliminate the need for a multitrack ATR when laying audio tracks against video. With AudioFile, editing may be performed within the digital domain; due to the fact that this editing is fully electronic and random access, it can be carried out on a single machine within an accuracy range of microseconds.

Fig. 2-35. AudioFile, a computer-based ATR system. (*Courtesy of Advanced Music Systems Industries.*)

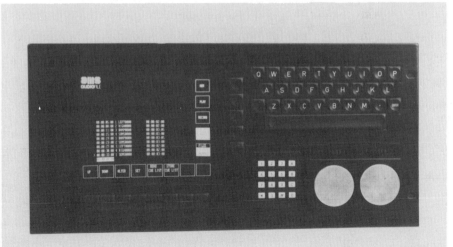

The Video Tape Recorder

In its basic operation, the *video tape recorder*, or *VTR*, is similar to the audio tape recorder in that the video signal is stored on magnetic tape for later reproduction. This simple comparison extends no further than that, however, since the recorded video signals are stored and processed in a different fashion than are the signals in audio recording. The most obvious reason for the difference in approach is the need to record a much greater signal bandwidth onto magnetic tape in video recording. The required bandwidth for a quality audio signal is roughly 30 Hz to 18 kHz, corresponding to a range of about 10 octaves; the bandwidth required for video spans an 18-octave range, from 30 Hz to 4.5 MHz. Achieving such a wide range using conventional audio recording methods would require conditions that would make video recording totally impractical.

Basics of Video Recording

In the video recording process, the entire 18-octave bandwidth is important to the overall signal. The upper set of frequencies covers picture quality, and the lower frequencies are required for sync pulses.

In magnetic recording, the major restriction governing the placement of video signals onto tape lies in the bandwidth limitations of the playback head. For the ATR reproduce head, the upper frequency cutoff point is determined by the width of the head gap; this consequence is known as the *gap effect*. As the signal frequency rises, a point is reached at which the entire recorded wavelength fits within the gap width of the reproduce head, resulting in total signal cancellation at the head's output (Fig. 2-36). The design of the ATR reproduce head is such that these limitations fall within acceptable tolerances. However, a gap width capable of reproducing the upper ranges required for video signals would need to be so minute that it would produce too small an output, resulting in a poor signal-to-noise ratio. A solution that might seem acceptable would be to increase the transport tape speed, thereby increasing the effective recorded wavelength and thus the bandwidth. This fails to be the best answer to the problem because excessive tape speeds lead to high tape costs as well as storage problems.

The greatest restriction on the reproduction of video signals is the frequency-dependence of the reproduce head. Since the signal output of a head inherently rises at a rate of 6 dB per octave, a video head designed for an optimum output of 4.5 MHz would reproduce the 30-Hz sync pulse at a level 110 dB below the point intended. From this, it is clear that the entire bandwidth for video signals is simply too wide a range to be recorded by conventional means.

The means adopted to circumvent this problem was a shifting of the bandwidth up to higher frequencies. By translating this spectrum upwards, to between 5 MHz and 10 MHz, the 5-MHz bandwidth is left intact, but the ratio of the highest frequency to the lowest becomes only 1 octave, instead of 18. With a 1-octave range, a reproduce head with a gap optimally designed for 7.5 MHz would reproduce the entire spectrum with ease, giving an optimum output level and a difference in level of only 6 dB at the bandwidth extremes.

The technique that allows the frequency spectrum to be shifted upward is *frequency modulation (FM)* of the video signal. Modulating the frequency of a carrier signal in direct relation to an incoming video signal yields many advantages. Since the demodulation of an FM signal works to remove all variations in recorded amplitude (so amplitude is not a direct factor in the recording process), the 6-dB variation

Fig. 2-36. Demonstration of the frequency-selective gap effect on the video signal.

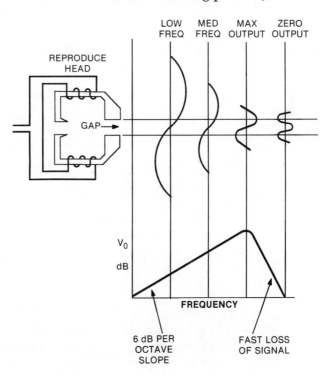

over the recorded frequency range is eliminated, as is the need for any form of playback equalization. The FM signal may thus be recorded on tape at a level approaching saturation, which ensures a strong signal level.

One other major advantage of FM recording is the insensitivity to dropouts that are due to erratic contact between the video heads and the tape. Sideband modulation noise and distortions are also kept at a minimum, which results in further improvements in the overall signal-to-noise ratio and the picture quality.

In practice, a high-frequency waveform of sufficient amplitude is used. The frequency is varied in direct proportion to the video signal, with the lowest carrier frequency representing the sync tips and the highest carrier frequencies corresponding to the peak white end of the spectrum (Fig. 2-37). Many VTR machines operate using a sine wave for the FM carrier, whose frequency can range, depending on the machine, from approximately 2 MHz to as high as 10 MHz. A few of the newer helical VTRs make use of a square wave for the FM carrier, as this type of signal is easier to modulate accurately.

Fig. 2-37. Frequency bandwidth of the conventional video signal.

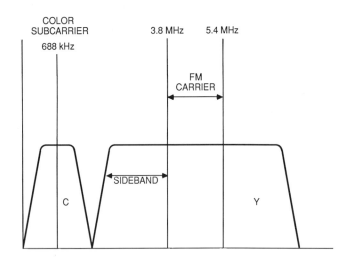

Many of the basic problems associated with the recording of video signals are solved by the use of frequency modulation to reduce the bandwidth from 18 octaves down to a more manageable spread of less than 2 octaves. Since the FM carrier is positioned in the megahertz range, a high tape speed and a small head gap are still necessary.

The Longitudinal VTR

The first experimental video tape recorder incorporated a conventional longitudinal tape transport. A transport speed of about 200 ips is the maximum that may be achieved with ease with such a mechanism. Given a head gap width of 0.1 mil and a tape speed of 200 ips, a bandwidth of only 1 MHz can be achieved. This figure would produce a picture that was totally inadequate for video broadcast purposes. To overcome this difficulty, a method of splitting the video signal equally across several tracks of a longitudinal tape was attempted, with considerable success.

In 1955, the BBC went on the air in London with a machine called VERA, which it had jointly developed with Decca. VERA operated by splitting the video bandwidth between two tracks. Track A contained the sync and video information from 30 Hz to 100 kHz, and track B was used for the video signal between 100 kHz and 3 MHz (the British 405-line video scanning system requires only a 3-MHz bandwidth for standard operation). These signals were then frequency-modulated at a carrier frequency of 1 MHz. At about the same time, Crosby Enterprises (owned by Bing Crosby), in close association with Ampex Corporation, was developing a longitudinal recorder. Both VERA and the Crosby/Ampex machine, however, had several major problems that could not be overcome, such as transport instability, a restricted bandwidth, high tape consumption, and (therefore) restricted recording time. Obviously, the use of longitudinal transports was not suitable for video broadcast recording, and some other technology had to be developed.

The Quad VTR

The solution to the transport problem was found in the *quadruplex*, or *quad*, *VTR*. This device combined the advantages of a high head-to-tape writing speed and a lower transport speed by incorporating a rotating magnetic head into its design. In 1957, the world's first commercial video tape recorder was introduced by the Ampex Corporation—the VR-1000 (Fig. 2-38). This machine recorded only black and white (but was soon available with a color option) and was designed to very tight industry specifications as a broadcast machine.

Fig. 2-38. Ampex VR-1000, the first commercial video tape recorder. (*Courtesy of Ampex Museum of Magnetic Recording, Redwood City, CA.*)

Head and Transport Assembly

In the quad VTR, the record/reproduce head assembly is composed of a rotating wheel on which four magnetic heads are mounted at 90° angles relative to one another (Fig. 2-39A). The heads are mounted at the periphery of this wheel and rotate across the width of the tape, scanning at a right angle to the direction of tape travel (Fig. 2-39B).

Since tape-to-head contact is very critical in video recording, the tape is held in contact with the rotating heads by a curved guide, which ensures precise vertical alignment and causes the tip of the recording head to penetrate slightly into the tape's magnetic oxide. This head penetration must be carefully controlled. If the tape tension is too great, the head will gouge out minute sections of the tape surface and will gradually wear away the magnetic oxide.

Fig. 2-39. The quadruplex (quad) video path.

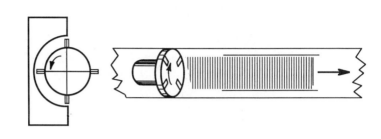

(A) The head scan path.   (B) The recorded video signal path.

The head wheel of the quad VTR is approximately 2 inches in diameter, with a total circumference of 6.28 inches. Operating at a tape transport speed of 15 ips, the head wheel is designed to rotate at 240 rpm, giving a total tape speed relative to the heads of 1507 ips. Since the speed of rotation is four times the video field rate, one revolution will put one quarter of a video field onto tape. Because a multiple head system is required to record a full video signal on tape, some method of switching from one head to the next must be employed. This is necessary for two reasons: to reduce the added noise that is generated by the unused heads and to prevent the overlap of signals where two heads are in simultaneous contact with the tape.

The Quad Recording System

In the record mode, the input to the quad VTR is made up of the *composite video signal* (Fig. 2-40), which contains all of the necessary video and sync pulse information. This signal is preemphasized at the upper frequencies in order to improve the signal-to-noise ratio and is then fed into the FM modulator. The modulated signal is distributed to all four heads, with no electronic or mechanical switching. There is a slight degree of head overlap as tape contact is made, so some duplication of the signal occurs between one head pass and the next.

The quad recording system is thus fairly straightforward. The tape transport and heads maintain a constant speed so as to place the recorded tracks on the tape in an orderly and controlled fashion.

In the record mode, the tape tension is highly regulated at the transport, to provide for smooth operation (a condition important to the process of video reproduction). At the same time, a recorded pulse is placed onto the videotape in a longitudinal direction; this longitudinal pulse is known as the *control track*. The control track, which is triggered from the sync pulse of the composite video signal, acts as a signal on the tape for timing and tape tension control in the reproduce mode.

Most problems besetting the quad video recording chain occur in the reproduce mode, as it is not a simple procedure to retrace the vertical lines of the video track accurately. In the record mode, it is a straightforward enough process to record the video signal on tape in a controlled manner. However, upon playback, the tape must be in exactly the same position at a precise time in order for the rotating heads to retrace the recorded video tracks. In addition, for broadcast applications, where the playback signal may be fed to or mixed with signals from other video devices, synchronization, or *video sync*, must be achieved housewide if switching from one video source to another or mixing is to be done without resulting in visible distortion.

Fig. 2-40. The composite video signal.

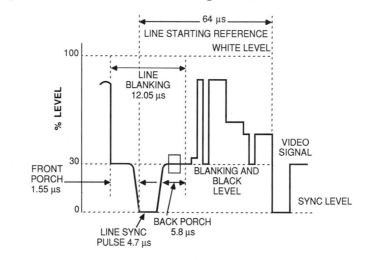

Head Switching and Transport Regulation

Since a definite pattern for the reproduction of a video signal from a tape is required, the rotating heads must precisely retrace the tracks as they were recorded. The tolerances for this retracing are quite tight; even minute tracking misalignments can result in loss of level or visible picture breakup due to crosstalk. To lessen the numerous problems involved in the reproduction process, precise locations along the head-to-tape path are chosen for the vertical interval and the head switching points within the video field.

In the record mode, as has been mentioned, a sync pulse (so named because it is locked to the sync tip contained within the vertical interval) is derived from the incoming composite video signal and is recorded onto a separate longitudinal track known as the control track.

In the reproduce mode, the video sync is derived from and thus is locked to this recorded control track. Tape speed, tape tension, tape positioning, and even electronic head switching are controlled by this extremely important pulse. Tape speed, tension, and positioning are regulated by a continuous comparison of the relative timings of the recorded signal's sync tip and that of the control track. Should the two signals not match, the transport is sped up or slowed down until the sync points do match, thus ensuring precision tape placement. Timing of the head switches is similarly regulated by the control track signal. This pulse is used to derive a 240-Hz squarewave signal, thus synchronizing the head wheel servo motor for exact positioning.

The Helical Scan VTR

In 1960, a new form of video tape recorder, known as the *helical scan*, or *helical*, *VTR*, made its way into the nonbroadcast market for use in closed circuit television systems. Although the reproduction quality of the quad VTR was unsurpassed, the machine was also extremely large, expensive, and difficult to maintain. These drawbacks, together with the advent of new applications and the solid-state transistor, created a market for a smaller, simpler, and less expensive unit.

As described above, the head of the quad VTR scans transversely across the length of the magnetic tape at a high rate of speed. Each recorded track contains 16 lines of video information and four complete revolutions of the head are required to complete one video frame. Thus, a highly sophisticated and accurate head-switching network is required in a quad VTR in order to record and reproduce the whole video field. A great simplification in design was envisioned if an entire video field could be recorded within a single pass, eliminating the need for critical head switching and phase matching.

This simplification was realized when a VTR using a helical scan tape path was developed. In the helical scan system, the tape is curved around a large rotating drum, containing one or two magnetic video heads that rotate in a plane parallel to the machine's transport deck (Fig. 2-41A). The tape wraps around the drum in such a way that it emerges from the tape guides at a different level from the one on which it entered (Fig. 2-41B). Because the tape describes a helical path as it moves around the drum, it is possible to achieve a very sharp angle between the track and the edge of the tape. This allows for a much longer recorded track. With the proper combination of drum diameter and tape speed, it is possible to record an entire video field on the tape within a single head scan.

**Fig. 2-41. The
1-inch, Type-C
helical scan
video path.**

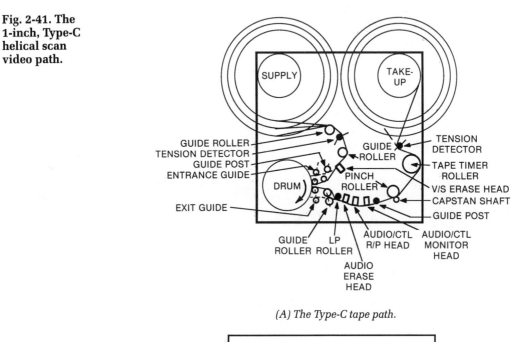

(A) The Type-C tape path.

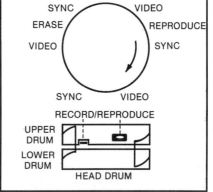

(B) The Type-C helical scan and wrap path.

As a result of these changes, the helical scan VTR is very different
in appearance from its quad counterpart. It is also lighter, cheaper, and
electronically simpler.

The Head Drum

The most critical component of the helical scan VTR is the head drum
assembly. This assembly houses a disc on which the video magnetic
heads are mounted. This disc is rotated in a horizontal plane within

the drum, while the tape is made to travel in a slanted path around the drum's perimeter (Fig. 2-42A). The resultant recorded tracks are therefore laid out in a parallel series at a high tape-writing speed. As the VTR transport moves the magnetic tape along at a constant rate, the rotating head disc cuts a series of long scans at a sharp angle to the edge of the tape (Fig. 2-42B). As was previously emphasized, in the reproduction of a video signal, it is absolutely necessary that the recorded track be scanned as accurately as possible. Because the helical scan VTR describes the recorded track at a sharp angle with respect to the tape edge, accuracy and transport control are critical.

Fig. 2-42. Relationship of head drum, tape path, and slanted tracks in the helical VTR.

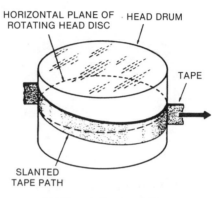

(A) Tape path around drum.

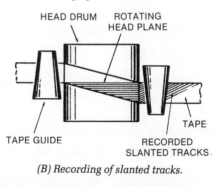

(B) Recording of slanted tracks.

The character of the tape path separates helical scan VTRs into two types: the *one-head, full-wrap VTR* and the *two-head, half-wrap VTR*.

The One-Head, Full-Wrap VTR—Many of today's professional helical scan VTRs are equipped with only one video record/reproduce head. With such a system, an entire video field may be scanned without any need for switching between alternating heads. This means that the rotating head must maintain constant contact with the magnetic tape during the scan of the video field. In order to accomplish this, a full wrap is employed, in which the tape path completely encircles the head drum except for a small section that allows for tape path entry and exit points. In tape drives using the *Omega full-head wrap* (Fig. 2-43), the head begins its scan at some small distance away from the edge of the tape, thereby leaving some unrecorded width for audio, cue, and control tracks. Since one complete video field is recorded during each head scan, the time required for the head to make a complete rotation is equal to the time for recording one field, which is 1/60 second.

Fig. 2-43. The Omega full-head wrap used in professional Type-C VTRs.

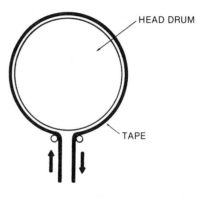

With a full-wrap system, a small portion of the signal is inevitably lost as the head passes from the end of one tape scan to the beginning of the next. However, this gap is purposefully located within the vertical blanking interval, where it is totally away from the visible picture area. As a result, a small dropout will occur within the vertical blanking interval.

Refusing to accept this small dropout, Sony Corporation decided to adopt a system in which an extra head was placed slightly in advance of the main signal head to describe a shorter and separate helical path for the recovery of the lost interval information. This system came to be known as the *1.5-head system*. It was later adopted for the *Type-C format* (Fig. 2-44), which has become accepted as a standardized compromise between the Ampex and Sony 1-inch formats.

Two examples of helical scan VTRs utilizing the Type-C format are the Ampex VPR-3 (Fig. 2-45) and Sony's BVH-2800 (Fig. 2-46).

The Two-Head Half-Wrap VTR—Another class of video recording system, which has achieved popularity in industrial and educational applications, is the two-head, half-wrap VTR (Fig. 2-47). By incorporating two video heads spaced 180° apart on a rotating disc, a constant head-to-tape contact can be maintained with a tape wrap of less than one full turn. A tape wrap of slightly more than 180° is employed in many of these VTRs, so there is a brief moment when both heads receive information simultaneously. This means that a certain amount of head switching must take place in the reproduce mode.

Fig. 2-44. The 1-inch, Type-C format for video-tape, showing placement of audio tracks.

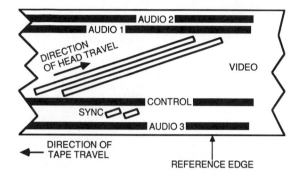

With a half-wrap system, each successive half-turn scan is used to record one video frame. Head switching is then set to occur just a few lines before the vertical sync interval, thus assuring that this switching point will not be seen on most video screens. Since the entire video field is scanned in half a rotation of the head wheel (1/60 second), an entire rotation is made in 1/30 second, which corresponds to half the rotation speed of a full-wrap system.

A presently accepted videotape format for half-wrap broadcast VTRs is the *Type-B format* (Fig. 2-48). One machine that utilizes this format is the Bosch BCN-51 VTR (Fig. 2-49).

The Helical Scan Guide Path

In the helical scan VTR, the tape is fed from the supply reel into the head drum at one level, from which it is made to follow a standard slant-wrap path around the perimeter of the drum. The tape then exits from the drum at a different level from that at which it entered and is fed onto the take-up reel. The direction of the helical path (either ascending or descending) varies from one manufacturer to the next and has no effect on performance, although the majority of professional VTR machines operate with an ascending helix.

In order to achieve this slanting path around the head drum, the tape is made to pass over a set of very finely machined guideposts (Fig. 2-50), which are located on both sides of the head drum. These guides may be skewed at an angle or constructed in the shape of a cone in order to offset the tape path. The ability of a VTR to maintain a constant tape path and thus to track the video signal accurately is directly related to the precision of these guideposts. In no other type of magnetic recorder is such a high degree of accuracy required; in addition, these guides must match precisely on all standardized VTRs or else compatibility would not exist.

Fig. 2-45. The Ampex VPR-3 Type-C VTR. (*Courtesy of Ampex Corporation.*)

Fig. 2-46. Sony BVH-2800 1-inch, Type-C, helical scan VTR. In addition to being fully compatible with all Type-C video and audio formats, the BVH-2800 offers two PCM channels with full digital audio specifications. (*Courtesy of Sony of America, Inc.*)

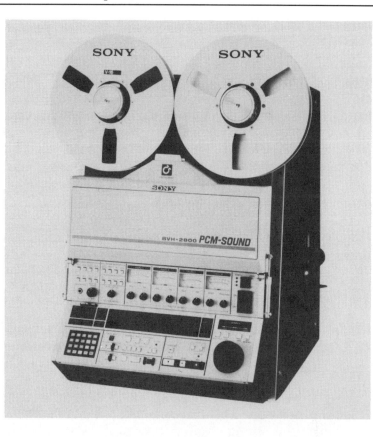

Fig. 2-47. Head wheel of a two-head, half-wrap VTR system. (*Courtesy of Robert Bosch GMBH.*)

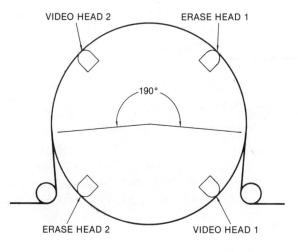

VIDEO HEAD 2   ERASE HEAD 1

190°

ERASE HEAD 2   VIDEO HEAD 1

**Fig. 2-48. The
Type-B format
for videotape.**
(*Courtesy of Robert
Bosch GMBH.*)

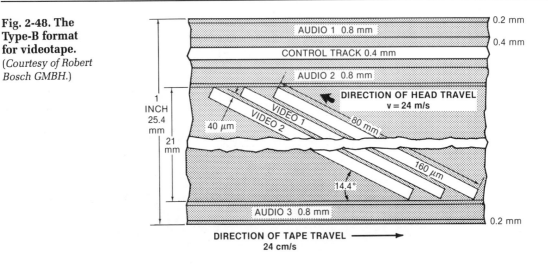

Fig. 2-48. The Type-B format for videotape. (*Courtesy of Robert Bosch GMBH.*)

The Magnetic Head

In principle, the video record/reproduce head operates much like its audio counterpart. However, it is quite different in design and bandwidth specification.

The head of a modern VTR is constructed out of a small chip of ferrite material (Fig. 2-51), with a thickness of about 10 mils (0.01 inch) and a gap width ranging from 40 to 50 microinches. Only six to eight turns of microscopic thickness are used in the pick-up coils of the video head.

In the video recording process, positioning of the magnetic head is critical, resulting in a need for precision head-mounting techniques. The small chip of ferrite is fastened onto a precision mounting plate, which is then used to mount the heads onto the rotating disc. Since mechanical tolerances in broadcast-quality VTRs are very strict, minor inaccuracies may be corrected by electronic pulse shaping and head servo regulation.

Rotary Head Connections—Because video heads rotate in a horizontal plane on a precision disc, some means of providing constant electrical contact is required. This requirement is met through the use of slip rings or a rotary transformer.

The set of slip rings (Fig. 2-52A) is usually mounted on the central post of the rotating disc, with wire leaf springs providing the electrical contact with the video electronics. The rotary transformer (Fig. 2-52B) is similarly mounted on the head assembly center post, with one set of coils within the rotating disc assembly.

Fig. 2-49. Bosch BCN-51 Type-B VTR in console. (*Courtesy of Robert Bosch GMBH.*)

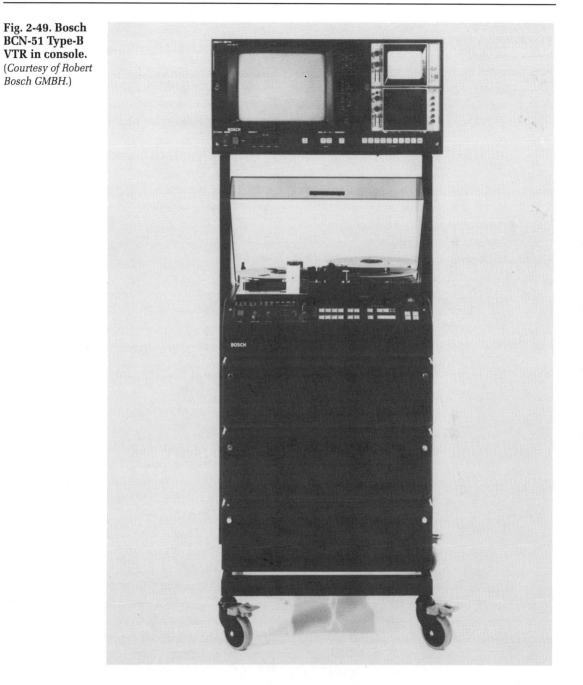

Fig. 2-50. Bosch Type-B head drum and guide path assembly. (*Courtesy of Robert Bosch GMBH.*)

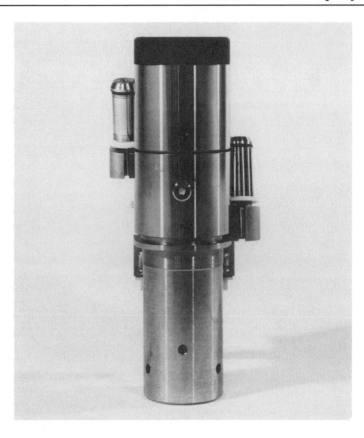

Head Degaussing—Like all magnetic recording heads, the video head will retain a small amount of remanent magnetization over periods of constant use. This internal flux may create a certain amount of noise in the video signal and may cause errors in the recorded color video signal. Therefore, it is important to degauss the video head periodically with a demagnetizer.

Demagnetizing the video head must be done with extreme care. The ferrite chip can actually be shattered if the head is touched directly with the demagnetizer or if the demagnetizer that is used is too powerful.

Fig. 2-51. The record/ reproduce head of a VTR.

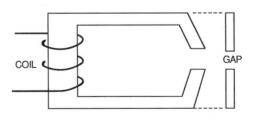

Tracking

In the record mode, the video tracks are laid down on the tape in as constant and controlled a fashion as possible. The tape drive and rotating heads operate at a constant speed without the need for control circuitry. However, in the reproduce mode, the process is much more involved. The reproduce head must follow the previously recorded track precisely if a video signal free from visual distortion is to be achieved. This retracing of the recorded signal in playback is known as *tracking*.

Electronics

The electronics of a helical scan VTR can be divided into two independent parts: the *servo control circuit* and the *record/reproduce circuit*.

In the record mode, the helical VTR puts a control track signal onto a separate longitudinal track, using as its source the vertical interval

Fig. 2-52. Mechanisms for achieving rotary head connection with VTR electronics. *(From Video Tape Recorders, Second Edition, by Harry Kybett.)*

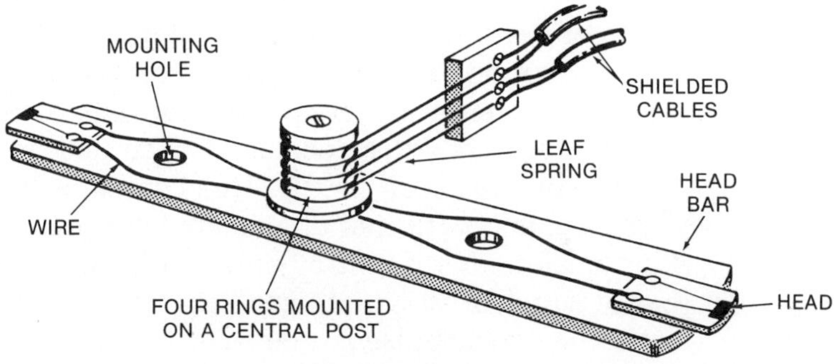

(A) Slip rings.

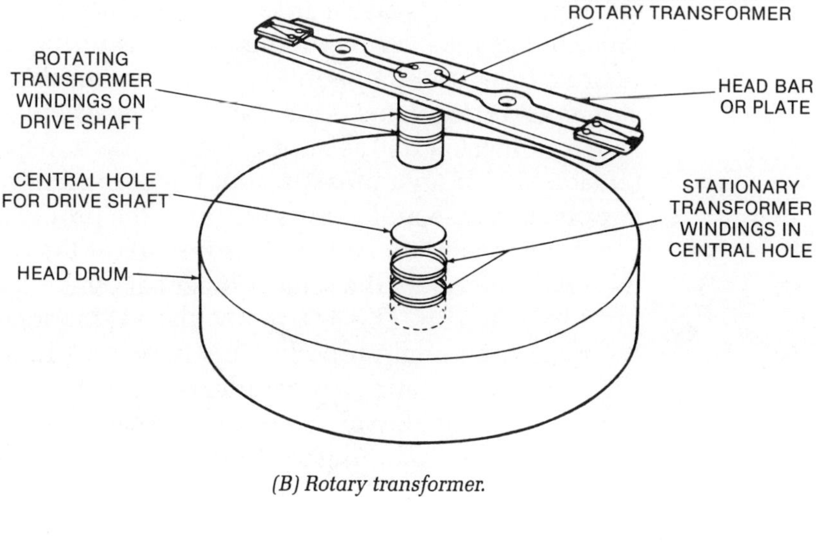

(B) Rotary transformer.

sync pulse. Thus, a reference is placed on the tape that will indicate the precise positioning of the video frame. On playback, this control signal is vital; it serves as a starting reference point for the placement of a video head scan.

Since control over tape and head scan placement is derived from a signal recorded on the tape itself, an interactive servo mechanism must be employed to control tape and head speed. Harry Kybett, in his book *Video Tape Recorders*, states that a servo motor and its associated circuitry must accomplish three things:

1. Place the heads in the correct position with respect to the tape and the incoming signal to record the signal in a definite pattern.
2. Effect correct tracking between tape and heads in the playback mode.
3. Effect head switching during both recording and playback.

In order to ensure proper transport speed and tape tension in the reproduce mode, a form of transport feedback is applied through servo control in order to achieve proper tracking. A constant tape speed that is synchronized with respect to the recorded signal is attained through the use of a resolver circuit. This compares the previously recorded control signal with that of a tachometer pulse that is derived from a tape tension idler (Fig. 2-53). Should the transport speed deviate from that at which the original signal was recorded, the resolver will sense a difference in the pulse rate between the tachometer (actual tape speed) and the control track (recorded signal speed) and will correct for the discrepancy.

Tape tension is also a critical factor since a high degree of head-to-tape contact must be maintained while avoiding stretching of the tape. In order to assure compatibility with other machines, a VTR's back tension must be regulated. This is most often accomplished by a tape tension arm located on the supply side of the transport. Should the tape tension vary from a preset amount, a corrective voltage is sent to the supply turntable motor, thus restoring the proper tension.

In a helical scan system, it is important that the video head begin its scan at precisely the same moment at which the recorded track enters the scan path. This is a prerequisite if the rotating head is to track the recorded signal properly. This important function can be accomplished by designating a particular place on the tape's scan path to contain the vertical interval. In the helical scan format, the vertical interval is placed at the beginning of a tape scan (Fig. 2-54). In most Omega full-

Fig. 2-53. Basic diagram of the Sony BVH-2000 servo system. (*Courtesy of Sony Corporation of America, Inc.*)

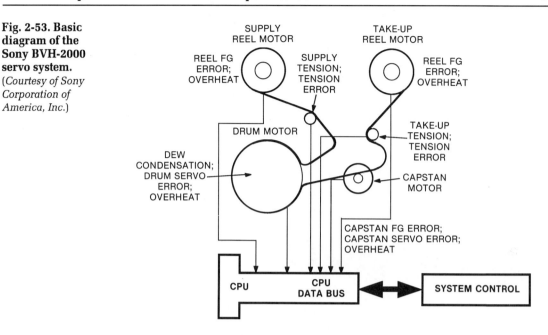

Fig. 2-53. Basic diagram of the Sony BVH-2000 servo system. (*Courtesy of Sony Corporation of America, Inc.*)

head wrap machines, the head is designed to come into contact with the tape at the beginning of the vertical blanking interval and is switched into circuit at three to five lines before the sync pulse.

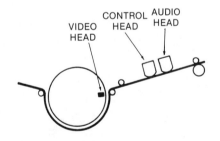

Fig. 2-54. Head and tape position for placement of vertical interval. (*From Video Tape Recorders, Second Edition, by Harry Kybett.*)

The Recording Chain—The recording chain of a helical scan VTR is quite simple in design. This chain consists of an input video amplifier, an FM modulator, and a record head amplifier.

In essence, the video signal is boosted by the input amplifier, limited in level by an automatic gain control (AGC) device, and subjected to a high-frequency preemphasis, in order to reduce the effects of tape noise on the signal. At this point, the signal is frequency-modulated with a sine wave or a square wave pulse (which is stable in waveshape) in order to shift the bandwidth upward. After modulation, the signal is fed to the record head amplifiers and the heads themselves.

In the reproduce mode, the output from the heads is a very low level FM signal, which must be preamplified. The amplifiers that accomplish this are often found within the top half of the head wheel itself. The boosted output is fed to a limiter, which effectively eliminates any tape-induced amplitude fluctuations, and is then amplified once more and routed to the FM demodulator. Once it has been demodulated, the signal must be deemphasized, to restore the high-end spectrum balance, and finally applied to the output of the VTR.

The Electronics-to-Electronics Mode—The *electronics-to-electronics (E-E) mode* is a monitoring facility peculiar to the professional video tape recorder. In the E-E mode, the monitor output of the VTR is allowed to pass through all the electronic stages of record-modulation and reproduce-demodulation. This allows for a quality check of the entire recording system while in the record mode. The performance of VTRs not equipped with the E-E mode may only be checked on playback of the recording.

Development of the Professional Helical Scan VTR
The first commercial video tape recorder was introduced by the Ampex Corporation in 1957; a fact that is not so well-known is that helical scan recorders were also in development at the same time as this transverse (quad) project. The quad system was ultimately chosen over the helical one because transverse recorders, although complex and expensive, were very forgiving of many types of mechanical problems. Helical scan recorders, on the other hand, were perceived as simpler, capable of certain tricks (like slow motion and still frame), and often less expensive. Unfortunately, the helical VTR had two inherent weaknesses which in the 1950s seemed insurmountable. The first was that accurate tracking of the recorded signal over the long slant angle was tricky. The second difficulty encountered with the early helical slant track recorder was an excessive amount of time base errors.

Slow Speed and Still Frame—Before the advent of the helical scan VTR, the functions of *slow speed* and *still frame* had not been implemented with any degree of success. The principle behind these modes of operation is based on the ability of a VTR to make multiple scans of a single video field. In slow speed operation, repeated scans are made of a single field as the tape is slowly shuttled over the head drum in the forward or reverse direction. To achieve a still frame, the tape transport is brought to a complete stop, while the head continues to scan the stationary video field.

In either of these modes, the orientation of the head with respect to the tape path is altered due to the varying or reduced speed of the tape as it passes the head (Fig. 2-55). As a result of this changed geometry, it is not possible to scan the recorded track accurately without introducing some type of compensation. Since the recorded FM signal can be processed to smooth out differences in level, tracking dropouts are not a critical problem. With certain helical VTRs, however, as the head travels into the guardband, a *noise bar* may appear. Also, mistracking will result in a lengthened path of the head across the tape, and the vertical sync intervals become spaced further apart. Since the tape speed is not constant in the shuttle mode, the timing between these intervals may vary. This can cause a variation in the triggering of the vertical stage on a video monitor, resulting in an unstable picture that may tend to bounce.

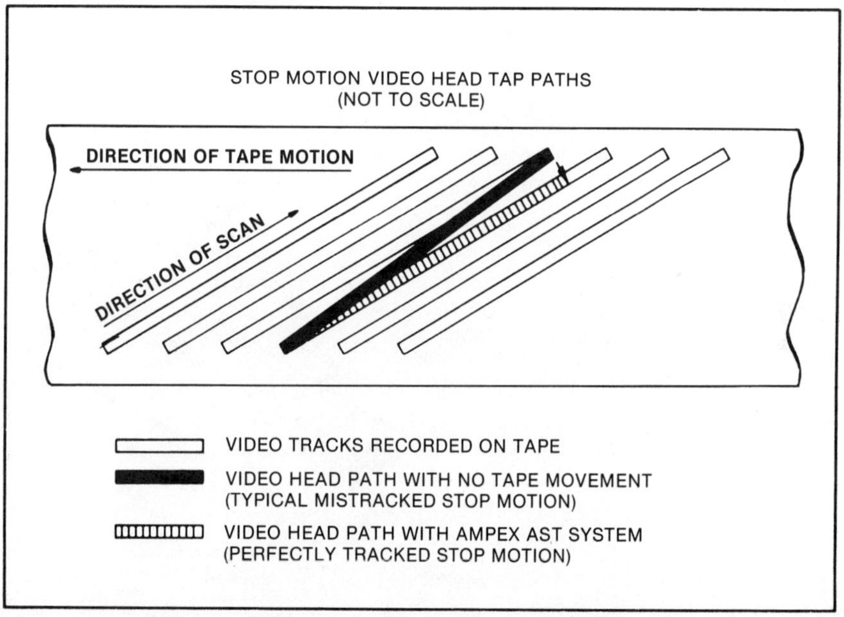

Fig. 2-55. Diagram showing tracking errors due to slow or still frame motion within a helical scan VTR. (*Courtesy of Ampex Corporation.*)

STOP MOTION VIDEO HEAD TAP PATHS
(NOT TO SCALE)

DIRECTION OF TAPE MOTION

DIRECTION OF SCAN

VIDEO TRACKS RECORDED ON TAPE
VIDEO HEAD PATH WITH NO TAPE MOVEMENT
(TYPICAL MISTRACKED STOP MOTION)
VIDEO HEAD PATH WITH AMPEX AST SYSTEM
(PERFECTLY TRACKED STOP MOTION)

To prevent this condition, two types of compensation have been used. Many professional helical VTRs correct for time-based errors by inserting an artificial vertical interval pulse into the video signal. This is inserted just before the final output stage and thereby serves to stabilize the signal. Another compensation system, developed by the Ampex Corporation in the early 1970s, is known as *automatic scan tracking*, or *AST*. In this system, the video head is able to correct for errors caused by mistracking because the head itself is mounted directly onto a piezoelectric transducer plate (Fig. 2-56). The incorporation of such a mounting plate makes it possible to deflect the tape dynamically in an up-and-down or articulated (azimuth) direction during the reproduce mode. With AST, the dynamic tracking information is obtained from the recorded signal itself. All necessary tracking information can be obtained by sensing the envelope of the reproduced video signal at the head. The dimension of this envelope is obtained by dithering the head physically from side to side along the recorded track at a high rate (450 Hz) to detect the amplitude and phase errors of the reproduced signal. Then error correction in the tracking of slow motion or still frame can be applied.

Fig. 2-56. The Ampex AST video playback head. (*Courtesy of Ampex Corporation.*)

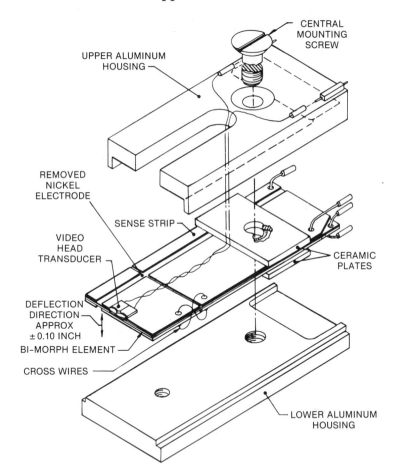

Time Base Correction—It has only been in recent years that the helical VTR has gained wide application in professional broadcasting and video production. The main reason behind this lag in acceptance is the fact that during the reproduction mode both the video signal and the sync pulse contain a high degree of *timing errors*, as compared to the level for a stable sync generator used to supply house sync.

In modern video production, it is desirable to be able to use prerecorded material. However, in order for prerecorded material to be mixed in with live studio material or with material from other VTRs (as in the video editing process), the prerecorded signal, as played back from the VTR source, must be stable with absolute accuracy. Otherwise, the reproduced signal will be virtually unusable for mixing with any stable video signal.

Generally, the reproduced signal from a VTR is *not* stable; it contains wide variations in timing between sync pulses and FM modulation. These signal variations may be traced to irregularities in the tape path (wow and flutter), head motion, and tape-transport skewing. Such variations are known as *time base errors*. Before the helical VTR could be used for any serious video production work, such errors had to be corrected.

In order to solve this problem, a digital device known as a *time base corrector*, or *TBC*, was created. The time base corrector is able to correct for a wide degree of timing errors and to produce a broadcast-quality signal that may be mixed with that from any other video source. The advent of this device transformed the helical VTR into the standard broadcast machine it is today.

The basic operation of the modern TBC is as shown in the block diagram of Fig. 2-57. The incoming video signal is transformed into a digitized signal, and the resulting stream of information is then placed into a computer-type memory for short-term storage, without any signal degradation.

Fig. 2-57. Block diagram showing the basic operation of the time base corrector.

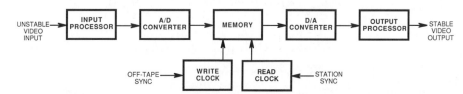

The sampling clock frequency, which governs the writing of the newly digitized signal into memory, is locked to the timing signal of the VTR in an "off-tape" fashion. Therefore, the information in memory contains all of the timing errors that were available at the output of the VTR. However, once the signal is placed into memory, it may be read out at a very stable clocking rate that is locked to the house video-sync generator. The result is a fully stable and synchronized output video signal.

Additional features, such as dropout compensation, picture enhancement, and picture stability in all shuttle modes, are available with many TBCs.

Color Picture Framing

In a monochrome video signal, a single picture frame is divided into two fields. Field one (the even field) begins with a full line and ends with a half line, and field two (the odd field) begins with a half line and completes its cycle with a full line (that is, odd-even-odd, etc.).

In the American NTSC color video signal, the subcarrier of each successive video line is phase-shifted by 180°. Therefore, the frames are oriented in a four-field sequence (that is, odd-odd-even-even, etc.), and once in every such cycle the field horizontal-sync pulse and subcarrier phase match up. With the European PAL (Phase Alternated Line) system, the phase of the color signal is reversed on each alternate line, so that this system employs an eight-field sequence for determining the color composition of a video frame.

The Video Cassette Recorder

Over the course of the last decade the *video cassette recorder*, or *VCR*, has gained prominence in both professional and consumer markets. One professional VCR is the JVC CR-850U (Fig. 2-58). In the professional market, the accepted VCR format has become the 3/4-inch U-Matic. In the consumer market, the battle among Betamax, VHS, 8-mm, and perhaps other formats yet to come has led to intense competition and innovation.

In principle, the video cassette recorder is very similar to the open-reel VTR, with the major differences being in appearance and format. The major difference in format is that the tape used with VCRs is enclosed within a magazine, or cassette, eliminating the necessity for reels. VCR units are generally compact in size, in many cases even portable, and some are constructed as VCR/camera combinations.

Since the tape is contained within a sealed cassette, tape threading must be done within the machine itself. The head drum of a U-Matic VCR is located just below the cassette when it is in the loaded position, and threading is accomplished by means of a large threading ring that describes a path around the outside of the drum, where the tape is supported by tape guides (Fig. 2-59). This threading ring is placed at a slant, in order to guide the tape into its helical tape path. A set of

Fig. 2-58. JVC CR-850U 3/4-inch U-Matic VCR. (*Courtesy of JVC Company of America.*)

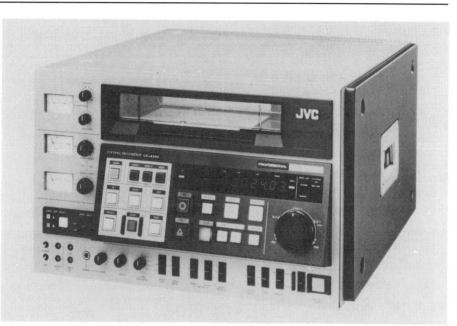

Fig. 2-59. Tape path within 3/4-inch U-Matic VCR. (*Courtesy of JVC Company of America.*)

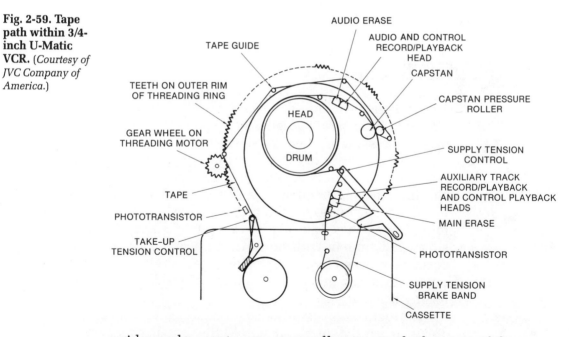

guides and a capstan pressure roller are attached as part of the transport and guide system.

With the newer Type-2 U-Matic VCR, it is possible to shuttle the tape in a forward or reverse direction with the tape left in a partially threaded path around the head drum and along the audio recording path (Fig. 2-60). Contact is maintained with the audio control tracks so that the control pulses and time code can be read while the VCR is in the shuttle modes. Partial visual cueing is also possible within these modes for use during editing or postproduction.

Fig. 2-60. Tape path of 3/4-inch U-Matic VCR equipped for fast forward and rewind with picture. (*Courtesy of JVC Company of America.*)

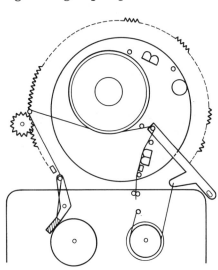

Digital Video Technology

The digitization of a video signal is accomplished with techniques similar to those used in digital audio processing. In the digital recording or processing of video, the signal with its 5-MHz bandwidth is converted into a series of discrete pulses, which are represented by the binary digits 1 and 0. As the incoming analog signal is sampled at a specific clocking frequency, it is converted into a series of digital *words*, which represent the precise value of the signal in encoded form. Once it has been placed in digital memory, the signal may be stored indefinitely with no resulting degradation. On playback, the signal is retrieved from memory at a specific clocking frequency (governing the speed of reproduced motion) and is reconverted back into its original analog form.

The digitization of the video signal was not practical until the early 1970s, when the development of LSI computer memories and high-speed analog-to-digital (A/D) converters helped to make it feasible. However, the digitization of the video signal requires a very fast and accurate conversion system, as well as high-density storage. Thus, video has been one of the last types of signals to which digital techniques have been applied, although advances are currently being made.

A few of the inherent advantages of digital video are as follows:

- Less susceptibility to noise and distortion
- Ease of transmission
- Ease of storage within digital memory
- Ease of manipulation using computer programs
- Ease of error correction and removal

In digital video, an 8-bit digital word is employed to encode the signal information rather than the 16-bit word required for digital audio. The analog video signal is frequency-modulated before encoding; the signal amplitude is not much of a factor in the digital video signal. This advantage is largely overshadowed by the fact that the sampling frequency required for normal playback speed must be higher than twice the highest subcarrier frequency. Thus, the necessary sampling frequency is very high, ranging from 10.7 MHz to 14.3 MHz, and to achieve this requires expensive and precise circuitry.

Digital Video and the Future

The use of digital video techniques in video production is presently growing at an escalating pace. Currently, digital video signals are being processed on a regular basis (with the time base corrector) in television studios and video production houses throughout the world. In addition, this technology has appeared in a multitude of new forms in the past decade.

Satellite transmissions of audio and video often utilize digital encoding techniques, providing station-to-station linkups with no loss of signal quality over long distances. Digital processing is also used in converting from one broadcast standard to another, such as from the American NTSC system to the European PAL or SECAM system, for international distribution.

Digital special effects have had an impact on both the professional and consumer markets. A digitized video signal can be manipulated and processed to suit any desired visual layout. One professional unit that can do this is the ADO™ 3000 system (Fig. 2-61), which is capable of executing such digital special effects as controllable aspects ratio (picture bend or inversion), controllable picture placement within a screen field, mosaics, and blur.

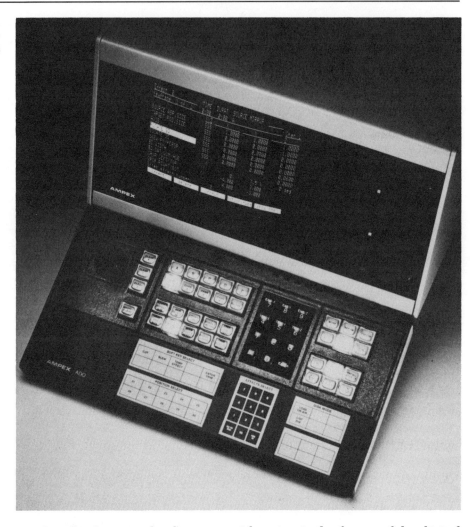

One final aspect that bears consideration is the future of the digital VTR. Although the development of the digital VTR is in its infancy stage (blocked mainly by the immense amount of memory required to store the digital video signal), one recent development on the digital video scene is the computer-based VTR, which is capable of storing a digital video signal directly on hard disks. One such machine is the A62 Digital Disk Recorder by Abekas Video Systems (Fig. 2-62). Billed as a special effects, animation, and editing tool, the A62 is capable of recording 50 seconds of composite video using one Winchester disk drive, or 100 seconds using two drives. When used in conjunction with just one source VTR and outboard special effects or editing devices, the A62 is able to create animation or unlimited multilayer special

Fig. 2-62. Abekas A62 Digital Disk Recorder. (*Courtesy of Abekas Video Systems, Inc.*)

effects without the generation losses associated with similar application of an analog VTR.

3 Synchronization: The Basic Building Block

The professional audio and video markets of today rely heavily on highly adaptive and flexible technologies. Current production methods utilizing combinations of film, video, and audio can be commonly found in the day-to-day operations of any one of the entertainment media. For example, postproduction techniques for video and multitrack audio have entered a new era in which multiple combinations of video and/or audio transports may be routinely operated together, in computer-controlled tandem, allowing the operator to refine a video edit or a recorded sound track to perfection. This degree of flexibility is the result of the ability to control numerous combinations of tape and film transports, which will function effectively together in synchronization.

Synchronization

Synchronization, often abbreviated as *sync*, is the occurrence of two or more events at precisely the same point in time. When used in reference to devices that hold audio or visual information (such as two or more tape transports), it means the ability to present information using multiple playback devices, such that recorded events are simultaneously reproduced. In the case in which several events are recorded on a single continuous piece of film or magnetic tape, the separate events are held in synchronization with respect to one another by being recorded on the same physical storage medium. The separate signals, as shown in Fig. 3-1, are recorded in a stable time relationship on the same base material and thus are physically locked in synchronization at all times. When more than one device is used to record or reproduce program material, synchronization is lost because more than one storage medium is involved. Recorded events are no longer held physically in sync and will vary in time relative to each other (Fig. 3-2). Even if two or more transports are started from precisely synchronized points, synchronization will soon be lost as a result of tape slippage and minute differences in tape speed. Thus it becomes clear that if production is to utilize multiple media and transports, some way of interlocking the multiple devices synchronously in time is essential.

Fig. 3-1. When separate events are recorded on different tracks of the same storage medium, they are locked into a physically synchronous relationship at all times.

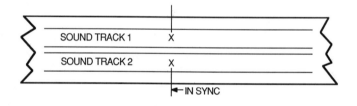

The assumption in the rest of this chapter is that the storage medium is video or audio magnetic tape, unless otherwise noted.

Fig. 3-2. Synchronization is lost when events are recorded on more than one physical storage medium since differences in speed and slippage will cause the relative relationship between the physical media to vary with time.

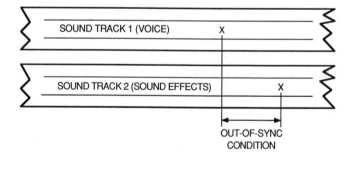

Phase-Locked Synchronization

One early way of assuring synchronous operation was a technique known as *phase-locked synchronization*. This method was based on the use of a *resolver* to maintain a constant transport speed. The resolver (Fig. 3-3) may be viewed as a comparator, whose function it is to compare the speed of a transport against a standard reference and to correct for any deviation. A sine wave that is known to be stable in frequency is usually chosen as the standard. Most often this sine wave is that of the frequency of the AC power line (60 Hz in the United States and 50 Hz in Europe) since it possesses great stability. At the time of recording, this standard frequency (the power having been lowered in level by passing the signal through a transformer) is recorded onto a spare audio track by the recording device; this track is thereafter referred to as the sync, or control, track. Upon playback, this control track, along with the standard reference signal of the same frequency, is fed into a resolver whose output controls the speed of rotation of the transport's capstan. Should any variation occur in transport speed, the frequency of the previously recorded control track will be proportionately altered. The resolver, by comparing this varying frequency with the standard nondeviating frequency, senses and computes any difference between them and will respond by changing the transport's motor speed until the frequencies are once again in relative sync (Fig. 3-4).

Fig. 3-3. Basic diagram of a phase-locked resolver system.

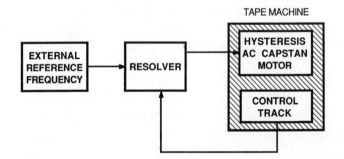

Phase-locked synchronization may also be used to interlock the speed of one transport with that of a second machine in a "master-slave" relationship (Fig. 3-5). In this configuration, the standard signal is recorded onto a spare audio track by each machine. Then one machine is designated as the master for playback (its control track signal becomes the standard reference) and the other machine is designated as the slave (its control track signal is compared with the standard). The resolver will correct for variations in the speed of the slave relative to the speed of the master transport. This type of phase-locked synchronization is a relative locking system. Whether it succeeds depends entirely on whether the two transports can be put into the play mode at precisely the same time with the two tapes at precisely the same location. Because of the cyclic or repetitive nature of the sine wave used for the sync signal, it is possible for the two transports to sync up on the wrong pulse, thereby becoming locked into an out-of-sync relationship (Fig. 3-6). Corrective adjustments to the transports' speeds must then be made manually.

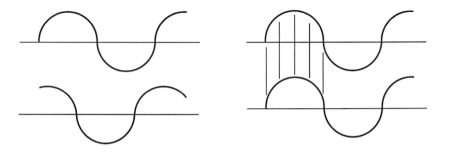

Fig. 3-4. A resolver circuit, through comparison of a reference frequency and that on a control track, is able to bring a transport into sync with the standard reference signal.

A fully detailed description of the system for phase-locked synchronization as put forth by Nagra/Kudelski may be found in Appendix B. The worldwide adoption of this amazingly simple system, known as Pilotone, by the broadcast and film media eventually led to the development of a great deal of equipment utilizing it. One such device is the EC-400 Universal Resolver (Fig. 3-7), developed by the Otari Corporation as a multipurpose ATR speed controller for use in applications involving the film-and-video interface. With the EC-400, the slave machine's control track may contain a Mono Pilot™ (biphase) or

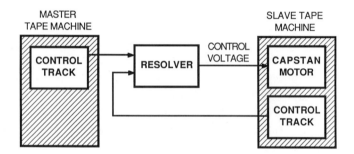

Fig. 3-5. Diagram showing resolver operating in a master-slave configuration.

an FM Pilot™ signal, SMPTE/EBU time code, or any signal in the range of 40–80 Hz. The external reference for the system may be obtained from SMPTE/EBU time code, a composite video signal, AC power lines, or any signal of 40–80 Hz operating at a TTL level.

Fig. 3-6. With phase-locked synchronization, it is possible for the reference signal and the control track signal to be out of sync by a full cycle. This offset condition may be corrected by manual adjustment.

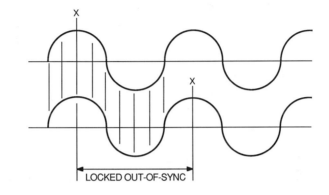

LOCKED OUT-OF-SYNC

Even though phase-locked synchronization has become universally accepted, there is a serious problem inherent in the phase-locking of material that has been recorded on magnetic tape. Phase-locked synchronization is only accurate in maintaining a regulated and relative transport speed. The resolver or controller is able to detect that the slave is out of sync relative to the master and to change the speed, but it cannot determine the actual tape location of either the master or the slave machine. To depend on such a system in modern postproduction work would be severely limiting. What is required is some system that allows both detection of variations in tape speed relative to a standard and detection of the exact position of the recorded tape at all times. Such a system would allow for precision synchronization and offset adjustments in tape position and would lend itself to the development of total computer control over position and function of multiple transports.

Time Code

Currently, the standard means of synchronization used in multimedia production work is *time code* (often abbreviated *TC*). The use of time

Fig. 3-7. Otari EC-401 Universal Resolver.
(Courtesy of Otari Corporation.)

code allows for the ready identification of an exact position at any point along a magnetic tape. A digital address code is assigned to each specific length of tape. This identifying address code cannot slip and always retains its original location, allowing for the continual monitoring of tape position to an accuracy of 1/30 second. The specified tape lengths consist of segments called *frames*, a term originating in film production. Each audio or video frame is tagged with a unique identifying number, known as the *time code address*. This address is usually an eight-digit number of the form 00:00:00:00, where the successive pairs of digits represent hours, minutes, seconds, and frames (Fig. 3-8). The time code address may also contain additional information, which will be discussed later in this chapter.

Fig. 3-8. Readout of the time code address takes the form HH:MM:SS:FF.

The recorded time code addresses work in much the same fashion for the location of a position on magnetic tape as postal addresses do for the delivery of mail. For example, if a mail carrier has a letter addressed to Ms. Smith at 47 Burnaby Street, he knows precisely where to deliver the letter because of the assigned house numbers (Fig. 3-9A). Similarly, time code addresses are used to locate specific positions on magnetic tape. For example, given a time-encoded videotape (Fig. 3-9B), which begins at 00:00:00:00 and ends at 00:28:19:05, assume that on this tape there is a specific cue point, a ball dropping, at time 00:15:07:26. Through monitoring of the address code (in a fast shuttle mode), it is a simple matter to locate the position corresponding to this address on the tape and then to perform whatever function is necessary at that specific point.

The Time Code Word

The electronic time-encoded information recorded for each audio or video frame is known as a *time code word*. Each word is divided into

Fig. 3-9. The location of a relative address.

(A) A postal address.

(B) A time code address on longitudinal tape.

80 equal segments called *bits,* which are numbered consecutively from 0 through 79. One word covers an entire audio or video frame, so that for every frame there is a corresponding time code address (Fig. 3-10). Address information is contained in the digital word as a series of bits

Fig. 3-10. The 80-bit time code word consisting of binary ones and zeros.

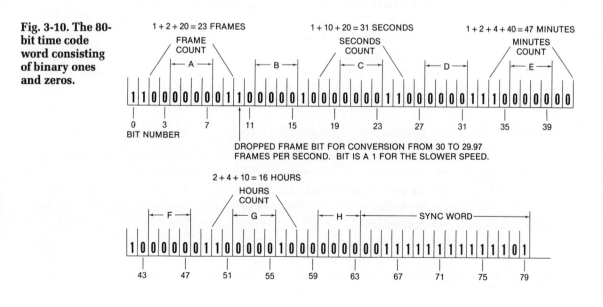

Fig. 3-11.
Biphase modula-
tion encoding.

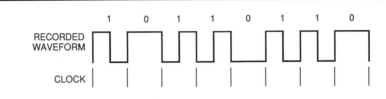

holding binary ones and zeros. These bits are electronically created and recreated to parallel fluctuations, or shifts, in the voltage level of the time code's data signal. This method of encoding information is known as *biphase modulation*. With biphase modulation, when the recorded signal pulse shifts either up or down only at the extremes of the time frame for a single bit, the pulse is coded as a binary zero, as shown in Fig. 3-11. A binary one is coded for a bit when a shift in the signal pulse occurs halfway through the bit period. A positive feature of this method of encoding is that detection relies on shifts within the pulse and not on pulse polarity. This means that time code may be read from magnetic tape in either the forward or the reverse direction and at slow or fast shuttle speed.

The 80-bit time code word is subdivided into groups of 4 bits (Fig. 3-12), with each grouping representing a specific piece of coded information. Within each of these 4-bit segments is the encoded representation of a decimal number ranging from 0 to 9 written in binary coded decimal, or BCD, notation. When a time code reader detects the pattern of ones and zeros within a 4-bit group, it interprets the BCD information as a single decimal number. Within the time code word, eight of these 4-bit groupings constitute the time code address in hours, minutes, seconds, and frames.

In addition to the encoded time code address, two other types of information are represented within the 80-bit time code word: user information and sync information.

User Bits
The 26 bits that represent the time code address are joined by an additional 32 bits called *user bits*. This additional encoded information, which is also represented in the form of an 8-digit number, has been set aside for users of time code to enter their own information. The standards committee of SMPTE (Society of Motion Picture and Television Engineers) has placed no restrictions on the use of this "slate code," which may contain such information as date of shooting, shot or take identification, reel number, etc. In addition to numerical information

Fig. 3-12.
Biphase repre-
sentation of the
SMPTE/EBU
time code word.

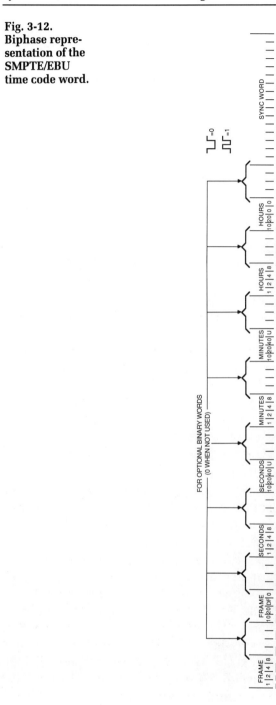

in BCD notation, more sophisticated systems allow for the coding of the letters A through F in hexadecimal notation (an 8-digit code might then be C1:A3:8F:B6). Currently, work is underway on a proposal of an expanded capacity for user bits, which would allow for the entry of related information in text form.

Sync Data

Another form of information encoded within the time code word is *sync data*. The sync data, found in 16 bits at the end of the time code word, are used to define the end of each frame. Because time code can be read in either direction, the sync data bits also function as a signal to the controlling device as to the direction in which the tape is moving.

Unassigned and Special Bits

The 26 time code address bits, 32 user bits, and 16 sync data bits add up to 74 bits, leaving 6 bits to complete the 80-bit time code word. These were originally unassigned bits intended for use with a future standard mode of operation. Four of these bits are still unassigned and are defined as permanent binary zeros. However, two bits have since been assigned a special function. A binary one in bit 10 or bit 11 indicates the use of drop frame code or color frame code, respectively.

Use of Time Code

Given the fact that a time code generator or reader measures in hours, minutes, seconds, and frames, the readout can be made to correspond with any desired time. The recorded code may be set to read out as the time of day, the elapsed time of the program material, or any other specific time that may facilitate multiple-reel productions. Within a given reel, time code makes address identification and logging quite easy since any given audio or video information is always at the same address, or calculated offset point. In multitransport or multimedia work, time code makes identification equally simple because of the fact that all recorded information at time X will be assigned the same address point on all machines. For example, if a ball dropping at 00:15:07:26 is to be previewed during a postproduction session using two videotape machines and an audio layback machine operating in sync, all that needs to be done is to simply set all the machines to that address (Fig. 3-13). The reading of time code from several transports is done by means of a central controlling device, thus giving the operator the ability to locate a precise event (to the frame) on all involved machines.

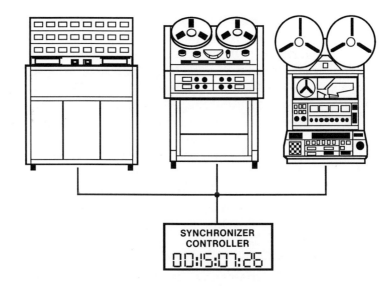

Fig. 3-13. A central synchronizer or edit controller can set several machines at a specific time code or tape location.

Nondrop and Drop Frame Code

The running of the time code for the counting off of its associated audio or video frames was, until fairly recently, directly related to the actual elapsed program time, an extremely important feature for the exacting time requirements of broadcasting. The black and white (monochrome) video signal operates at a rate of exactly 30 frames per second (fr/s). This monochrome rate is known as the *nondrop code*. If this time code is read, the time code display, program length, and the actual clock on the wall will all be in agreement.

This simplicity, however, was lost when the National Television Standards Committee set the frame rate for the color video signal at approximately 29.97 fr/s. This means that if a time code reader set up to read the monochrome rate of 30 fr/s is used to read a color videotape operating at 29.97 fr/s, the readout will pick up an extra 0.03 frame for every second that passes (30 − 29.97 = 0.03). Over the duration of an hour, the readout will differ from the actual tape address by a total of 108 frames, or 3.6 seconds.

In order to correct for this discrepancy and regain agreement between the time code readout and the actual elapsed time, a means of frame adjustment has been introduced into the code. Since the object is to drop 108 frames over the course of an hour, the code used for color video signals has come to be known as *drop frame code*.

Correction is accomplished by allowing the time code to advance as usual, except that the first two frame counts within each minute are dropped. The next frame following the one numbered 00:53:14:29 will be 00:54:00:02. The time code addresses 00:54:00:00 and 00:54:00:01 are omitted, although the actual video frames continue in an uninterrupted fashion. Since this pattern would account for the "loss" of 120 frames over the course of each hour and the object is to drop only 108 frames, some compromise is necessary. Therefore, drop frame code specifies the inclusion of the first two frame counts in every tenth minute. That is, in order to correct for a discrepancy of 108 frames per hour with the drop frame code, it is necessary to omit 2 frame counts for every minute of operation, with the exception of minutes 00, 10, 20, 30, 40, 50. This adjusts the frame count to agree with the actual elapsed time.

A binary one in bit 10 of the time code word is defined as indicating the presence of drop frame code, and a binary zero in bit 10 indicates the use of nondrop code.

Color Frame Code

Color video signals are not encoded on videotape in the same way as monochrome signals are. Due to the complexity of the encoded information, color video frames have an alternating nature, which can be simply indicated by designating them as either A or B. This alternating horizontal, or H, shift is created by the system's changing of the incoming video signal in order to create an A-B-A-B frame sequence. If an attempt is made to edit so that nonalternating frames are joined together (A-A or B-B), the videotape machine will display the error in the form of a temporary horizontal displacement of the signal at the edit point. An error in H shift is not noticeable if it occurs at an edit point between two different shots, because the small shift is obscured by the change in scene. However, if the mismatched frames occur within the same shot, the lack of exact match-up is clearly visible on the television screen.

To avoid visible H shifts, it is necessary to be able to identify the alternating video frames involved in an edit. Thus, each frame ending in an even number is defined as having the A characteristic, and each frame ending in an odd number is defined as having the B characteristic. The presence of a binary one in bit 11 of the time code word indicates the presence of this A-B color frame code.

The EBU Code

The *EBU (European Broadcast Union) code* in use throughout Europe utilizes an 80-bit time code word, as does the American standard. The only major difference is that the EBU code operates at a rate of 25 fr/s. Time code generators and readers must be able to adjust to this rate of 25 fr/s in order to work with EBU code and must be equipped to handle the European PAL 8-field color frame sequence if necessary. Since both monochrome and color video signals in EBU run at exactly 25 fr/s, there is no necessity for drop frame code or for any indication of its absence or presence within bit 10 of the time code word. With EBU code, a binary one in bit 11 denotes the presence of the PAL 8-field color frame sequence.

LTC and VITC

There are currently two major systems for encoding time code onto magnetic tape for broadcast and production use: LTC and VITC.

Time code that is recorded onto an audio track or a video cue track is known as *LTC*, or *longitudinal time code* (also commonly referred to as SMPTE time code). LTC is recorded onto a longitudinal audio or cue track (Fig. 3-14) in the form of pulse signals in a binary coded decimal (BCD) format.

Fig. 3-14. Otari MTR-10/12 center-track time-code option. *(Courtesy of Otari Corporation.)*

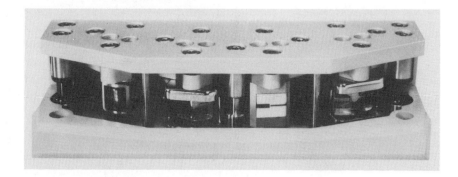

The rate of this binary pulse may be represented in bits per second (bit/s) or as a related frequency recorded as a square waveform. At the normal play speed of approximately 30 fr/s and with use of the 80-bit word for each frame, the result is a bit rate of 2400 bit/s, which is equivalent to a fundamental frequency of 1200 Hz. Since the bit rate is doubled in the recording of binary ones, the recorded frequency may likewise be doubled, to 2400 Hz.

The recording of a perfect square wave onto a magnetic audio track is, even under the best conditions, difficult. For this reason, SMPTE has set forth a standard allowable rise time of 25 ± 5 microseconds for the recording and reproduction of time code. This is equivalent to a signal bandwidth of 15 kHz, well within the range of most professional audio recording devices.

Variable-speed time code readers are able to decode time code at shuttle speeds ranging from 1/50 to 100 times the normal play speed. This range is effective for most applications; however, due to the limitations of magnetic recording, it is not effective at slower shuttle speeds or at still frame, modes that are often necessary for video editing. Since LTC cannot be read in the still frame mode, 3/4-inch and 1-inch videocassette worktapes, which are used for compiling edit lists for computer editing, must have the time code "burned into" the actual video image. The readout is often displayed inside a black box within the visible picture area, allowing the time code to be read in slow motion or still frame mode (Fig. 3-15).

Fig. 3-15. Video image with "burned-in" time code.

In video production, longitudinal time code has the additional disadvantage of taking up valuable space on the recorded videotape, space that might be put to better use for audio production work. In response to these drawbacks, an alternate system for the recording of time code onto videotape was developed. This code, called *VITC (vertical interval time code)*, utilizes the same structure for the time code word (address and user bits) as does LTC, but the signal is encoded on the videotape in an entirely different form. With VITC, the time code information is encoded within the video signal itself in a vertical strip, or field, located outside the visible picture area and known as the *vertical blanking interval* (Fig. 3-16). Since the time code information is encoded as part of the video signal itself, it is possible for 3/4-inch or 1-inch helical scan VTRs to read time code at slower shuttle speeds and in still frame. Also, there is no longer any need for the burned-in time code display of working videocassettes; instead, a time code reader is able to superimpose code that is frame-accurate onto the video screen when needed.

Fig. 3-16. Vertical blanking interval and sync.

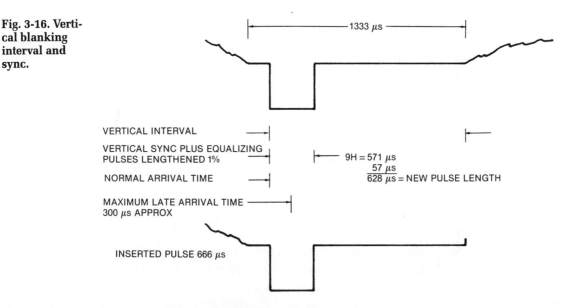

VERTICAL INTERVAL

VERTICAL SYNC PLUS EQUALIZING PULSES LENGTHENED 1%

NORMAL ARRIVAL TIME

MAXIMUM LATE ARRIVAL TIME 300 μs APPROX

INSERTED PULSE 666 μs

1333 μs

9H = 571 μs
57 μs
628 μs = NEW PULSE LENGTH

Another feature of VITC is that it can be used to identify individual fields, which are produced as the video system scans the odd lines (field 1) and even lines (field 2) of a video image. Since the code is spread over the entire frame, which comprises both video fields, it is a simple matter to identify the frame with field accuracy. This allows VITC to be accurate to 1/60 second during the video editing process.

VITC does have its disadvantages as well. For example, many videocassette transports are unable to read a video signal at all shuttle speeds, and 2-inch quad machines are unable to read a video signal at any speed other than normal play speed. In these cases, LTC must be employed. VITC also occupies vertical lines within the video signal area, which may sometimes be needed for user data. Finally, it is difficult to lay down VITC over a previously recorded video signal; in many cases, this requires that the signal be dubbed down a generation.

In most situations, LTC is preferred over VITC for production work. However, both forms may be utilized in order to gain the benefits of each in high-level video production work.

Shuttle Speed and Time Code

Assuming that the input level of the time code was above the minimum system requirements, difficulties in reading the code signal are most often a function of tape speed.

Code may be read at various tape shuttle speeds, with most readers being designed to decode the biphase signal in a range from 1/50 to 100 times the play speed. The physical limitations of the system are therefore related to the record/reproduce process.

In the reading of longitudinal time code, a minimum tape speed is imposed on the system. Typically, LTC cannot be read at speeds between 1/10 and 1/20 of play speed because the reproduce waveform tends to exhibit a high degree of distortion that can introduce extraneous or false data at the reader's decoding circuitry. The solution to this problem, in the production of video, is to utilize VITC, which is immune to errors at slow shuttle speeds and still frame.

The limitations on the reading of time code at high shuttle speeds are primarily related to the high-frequency response of the given recording system. At high speeds, the normal play speed bit rate of 2400 bit/s would be shifted upward into the 240K-bit/s range (for recording binary ones at shuttle speeds of 50 times the normal play speed). This is well above the frequency range of the average audio channel. Thus, a modification of a dedicated channel's electronics is necessary if it is to be able to handle such shifts in bandwidth. Many manufacturers of professional ATRs and VTRs presently offer modified-bandwidth record/reproduce cards as standard or optional equipment.

Jam Sync, or Restripping Time Code

With either LTC or VITC, time code is made operative by means of the recording of a series of square wave pulses onto magnetic tape. Unfortunately, it is somewhat difficult to record a square waveform onto analog magnetic tape without its suffering moderate to severe distortion. Although the binary-based time code reader is designed to be relatively tolerant of fluctuations in waveform amplitude, such distortion is severely compounded when code is dubbed down by one or more generations. For this reason, a special feature known as *jam sync* has been incorporated into most synchronizers.

It is basically the function of the jam sync facility to produce new code in order to match original time code addresses during the dubbing stage or to reconstruct defective sections of code. In the jam sync mode, the output of the generator is slaved to an external time code source. After reading this incoming signal, the generator will output a signal with the same time code address and user data information as the original code, but regenerated and free of distortions (Fig. 3-17).

Fig. 3-17. Representation of the recorded biphase signal.

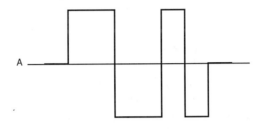

(A) Original biphase signal.

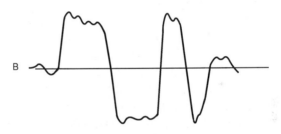

(B) Reproduced biphase signal.

One-Time and Continuous Jam Sync

Currently, there are two forms of jam sync in common use: one-time jam sync and continuous jam sync.

One-time jam sync is a mode in which, on receiving a valid time code address as the reader's initial input, the generator's output is initialized to that specific number and begins to count in an ascending order on its own, in a freewheeling fashion. Any deteriorations or discontinuities in the received code are ignored, and the generator will produce new, uninterrupted addresses as its output.

Continuous jam sync is used in cases where the original addresses must be kept intact. Once the reader is activated, the generator will update the address number for each frame in accordance with the incoming code.

Freewheel Mode

In the event of disturbances in the time code signal, which may be caused by tape dropouts or noise, the synchronizer may be prevented from making short-term changes in transport speed by being placed into the *freewheel mode*. This allows stretches of bad code to be passed over at a constant speed by all slave transports. In this mode, the synchronizer may make its calculations from short-term and long-term valid code readings.

Error Bypass

In addition to the above methods for improving or correcting time code, most modern time code readers have a special, built-in feature known as *error bypass*, which guards against the decoding of short-term false information. Error bypass is applied at the reader's output by a comparison of the actual decoded time code value with an expected value obtained from the preceding frame's address. If the next number does not match up with the expected value, the reader will output the expected value rather than the value that it read. This process will continue until some prespecified error limit has been reached, commonly two frames, after which the reader will again output the actual decoded time code value.

Although error bypass acts as an error-correction system for time code, it has the disadvantage of creating errors in one or two frames where a valid change or break in code occurs.

Plus-One-Frame

A time code reader receives a continuous stream of binary bits as its input, but will not update its output to the time code display until the end of a time code word has been reached, as indicated by the 16 bits of sync data. Thus, a reader does not display an address until after it has been read. To compensate for this delay, the reader must automatically add one frame to the frame count of the address currently being read in order for the real address and the display to match. This feature, known as *plus-one-frame*, is especially important during the scanning of video frames for editing purposes.

Displaying Time Code

There are currently three methods in common use for displaying time code to the production operator. The first is to display the frame-by-frame advancement of time code by using LEDs or LCDs located on the front panel of the time code reader or central control unit. A more recent trend with control synchronizers is to display the time code, along with all other information on the synchronizer's status, on a separate video monitor. In video production, it is often necessary to view the time code display and the video picture simultaneously. For this reason, most video time code readers have built-in character generators and inserters, which allow for the superposition of code on the video image. Both the size of the time code display and its placement within the video image may be selectable at the reader, but the displayed time code is made to stand out from the picture by placing it within a dark box located in the lower half of the video screen.

Synchronization Using Time Code

In order to achieve a frame-for-frame synchronous lock among multiple audio, video, and film transports utilizing time code, it is necessary to employ a device known as a *synchronizer* (Fig. 3-18). The use of LTC in audio production for video involves the synchronization of several ATR and VTR machines so that they operate in tandem as a single device.

During the last decade, synchronizers have become more and more sophisticated in their control over the operation and synchronization of multiple transports. Earlier synchronization packages were often composed of separate units for the generation, display, and synchronization of LTC. In the last few years, new types of multifunction synchronizers especially designed for use in audio production have come on the market. Although quite capable of audio/video synchronization, these devices are particularly useful for audio production work involving multiple transports.

The operation of synchronization systems has often been surrounded by misconceptions. The basic function of a synchronizer is to control one or more tape or film transports (designated as "slave" machines), whose starting position and tape speed are made to copy those of one specific transport (designated as the "master").

There are currently four types of synchronization systems that employ time code as their basis of operation: play-speed-only, chase, control, and edit decision list (EDL).

**Fig. 3-18. Audio
Kinetics Q-Lock
synchronizer.**
*(Courtesy of Audio
Kinetics, Inc.)*

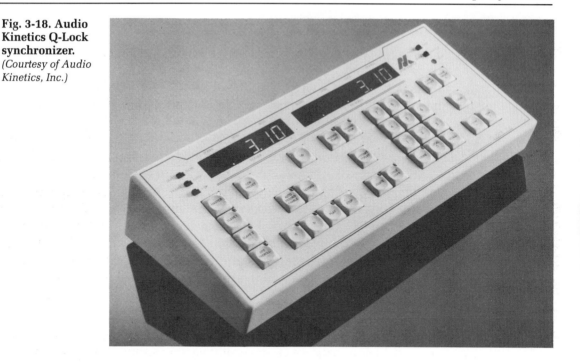

The Play-Speed-Only Synchronizer

The *play-speed-only synchronizer* requires that the master and slave
machines be manually cued closely enough that the slave machine
may be slipped into a synchronous lock by varying the capstan speed
(by 10% to 50%) until the timing error between transports is less than
one frame. Once frame lock is achieved, fine adjustments of less than
1% are made, and the involved transports are able to achieve and maintain sync by speed variations that do not produce audible effects.

The Chase Synchronization System

The *chase synchronization system* requires that a slave transport chase
the master under all conditions. This system is a bit more sophisticated
than the play-speed-only synchronizer in that the synchronizer has additional control over the slave transport's operating functions and is
often able to read the system time code for location cuing in the fast
shuttle or wind mode. This enhancement allows the slave transport to
switch automatically from the play mode into the cue (search) mode in
order to chase the master and then to resync when the master is again
placed into the play mode.

The Control Synchronization System

The *control synchronization system* places additional emphasis on flexible control of and interaction between multiple transports. The central control keyboard of such a system provides such options as machine selection, transport control, locate, looping, offset, event points, and record in/out, in addition to other features.

The machine selection feature allows the operator to choose the machine that is to be synchronized, as well as allowing for the selection of a designated master.

The transport control feature allows for conventional remote control over the functions of any or all machines in the system.

The locate feature is a command that causes the selected machines to locate automatically to a selected cue point by address.

The looping feature provides a continuously repeating cycle (play, repeat, play) between any two cue points, whose addresses may be stored in memory from the keypad.

The offset feature permits the correction of any difference in time code that may exist between coincident program material (that is, adjustment of relative frame rates by $\pm x$ frames to achieve or improve sync).

The event points feature allows a series of time code addresses to be entered into the control memory for use as cue points in triggering a series of function commands (for example, start slave or mastering machine, record in/out, and insert effects device or any externally triggered device).

The record in/out feature is an event function that allows the synchronizer to take control over a transport's record and edit operations, enabling in and out points to be located with frame accuracy.

The Adams-Smith System 2600 Modular Synchronization System (Fig. 3-19) was developed as a modular, time code, tape synchronization, and studio automation system. It offers such functions as LTC or VITC generating, reading, and translating, tape transport control, tape synchronization, audio editing, sprocket film sync, events control, computer interface, and character generation. The modules of System 2600 are able to communicate with one another by means of an internal, time-shared, multiconductor data bus. Each module is able to "broadcast" its address and data onto this common bus when a path becomes available, which means that modules can share or transmit data as needed. Messages consist of serial strings of one or more ASCII characters, which are distributed over a 34-conductor, flat-ribbon cable. These messages include commands that direct an interface either to receive data and broadcast it to a specific module or to pick up data from a specific module when it is broadcast and send it to the external communications equipment.

The mechanical design of System 2600 utilizes a patented "chassiRod" construction, with each module being a physically self-contained unit (including front and rear panels, I/O control unit, and connectors) so that any number of modules may be grouped together in any arrangement to form a specific network. This setup allows for changes and expansions to the system as well as ensuring compatibility of future with present modules.

Fig. 3-19. Adams-Smith System 2600 Modular Synchronization System. *(Courtesy of Adams-Smith.)*

A practical maximum of seven slave transports and a master may be controlled as a single system, using System 2600's dedicated keyboard controller or an external computer. System 2600 was specifically designed for ease of computer-based operation. Controls, indicators, and displays can be accessed externally over the data bus from either serial- or parallel-interface modules. This eliminates the need for large in-unit digital readouts.

Fig. 3-20. Adams-Smith Model 2600 CC, or compact controller. *(Courtesy of Adams-Smith.)*

Recently, Adams-Smith announced the Model 2600 CC, or compact controller (Fig. 3-20). This interface unit for System 2600 is a device small and light enough to be held in one hand and capable of controlling as many as five tape synchronizer modules and time code generators, thus implementing control and synchronization of combinations of up to five audio or video tape machines. The Model 2600 CC is equipped with fully automated multitransport audio rehearsing (preview), record, and playback, which are input into the system using time code addresses as in and out points. All necessary offsets, prerolls, postrolls, and durations are calculated automatically by the system. Additional features include eight nonvolatile function keys that permit users to store their own frequently used routines; a 100-position, 12-digit scratch pad memory; conversion between nondrop and drop frame codes; ATR vari-speed operation; and optional control (to an accuracy of 1/100 frame) over audio punch-in/punch-out or event switching.

The Edit Decision List (EDL) Controller/ Synchronizer

One recent entry into the field of audio/video synchronization is the *edit decision list (EDL) controller/synchronizer.* This development evolved out of the established use of the EDL within the video editing process. The EDL controller/synchronizer is, by nature, a device that is best utilized to control and synchronize various combinations of ATRs and VTRs for the laydown process of building tracks to a multitrack ATR from a series of discrete sources (such as multiple VTRs on which are recorded original field master audio tracks).

This system takes its name from the edit decision list that is compiled in the video editing process. In the course of creating a final edited master (EM) from the original field masters, a series of edits must be performed using a control edit computer. In order to accomplish these, a video editor has to exert total control over all of the involved VTRs, performing the necessary prerolls, synchronization, edit switching, and location of in and out points. This control is accomplished through the processing of a list of edits that are stored in memory as an edit data base. The control commands derived from this data base are input into the central controller computer in the form of a list utilizing time code reference points. This list consists of a series of precise reference points and special instructions as to tape positions, offsets, edit in and out points, etc.

The commands of the EDL serve to control and synchronize all of the associated devices within a video editing suite. The EDL controller/synchronizer's method of operation is quite similar to that of a control synchronization system, except that the EDL system is designed to be able to sync and switch among multiple VTRs, ATRs, and event-controlled devices (such as cartridge or random-accessed effects devices) in order to record on a multitrack ATR for audio postproduction. The use of the computerized EDL system provides the greatest degree of control and repeatability within the multitrack laydown phase of large-scale video productions.

CMX, Inc., manufacturer of the Edge EDL video controller, has recently introduced the CASS-1 (Computer-Aided Sound Sweetener), designed specifically for audio postproduction (Fig. 3-21). This device features the display monitor and color-coded keyboard that CMX has made familiar to many video editors. Machine control is achieved by an interface with the Adams-Smith System 2600, allowing up to six VTRs, ATRs, or 35-mm film dubbers to be accessed during the process. As many as fourteen event-controlled devices not regulated by time code may also be accessed via the EDL, providing a high degree of repeatability for nonsynchronous sources.

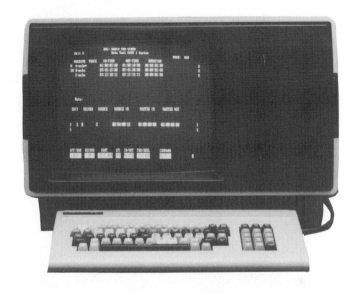

Fig. 3-21. CMX CASS-1, an EDL-based controller/ synchronizer. *(Courtesy of CMX, Inc.)*

When an operator selects material to be recorded on the multitrack ATR, all time code addresses for each edit are stored in the system's memory as part of the EDL. Additional information storage is provided by a 20-megabyte internal hard disk or an optional 8-inch CMX disk drive. The CASS-1 is also capable of accepting a standard video EDL from a CMX video controller, which may be found in many video editing suites. This capability makes it possible to load information from the video editing session directly into the CASS-1 for laydown.

The CASS-1 has been designed to operate in two different modes: the edit mode (for audio laydown under EDL control) and the mix automation mode (for VCA fader automation within a mix during the mixdown/layback phase).

Another EDL-based controller/synchronizer is Audio Kinetics Eclipse (Fig. 3-22). This "intelligent" unit is capable of synchronizing up to 32 machines, providing automated control during the audio edit or laydown phase of postproduction.

Fig. 3-22. Audio Kinetics Eclipse, an EDL-based controller/ synchronizer. *(Courtesy of Audio Kinetics, Inc.)*

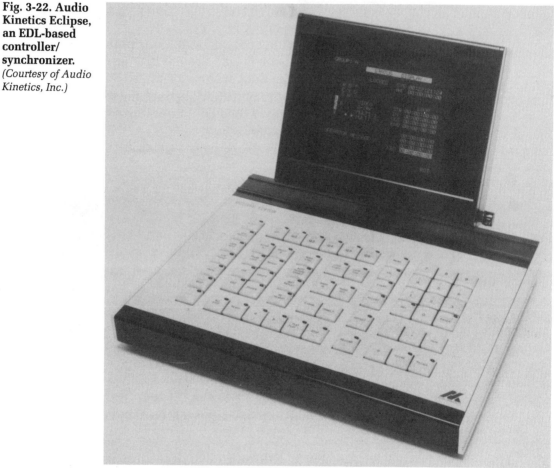

The Edit Decision List

The following is a brief summary of the format of the edit decision list (EDL) as it is utilized in the CMX CASS-1 for audio production for video.

The Edit Screen

The edit screen of the CASS-1 (Fig. 3-23) is divided into five sections:

1. Detailed edit area
2. Machine status area
3. Edit command and status area
4. Edit summary area
5. Mode display area

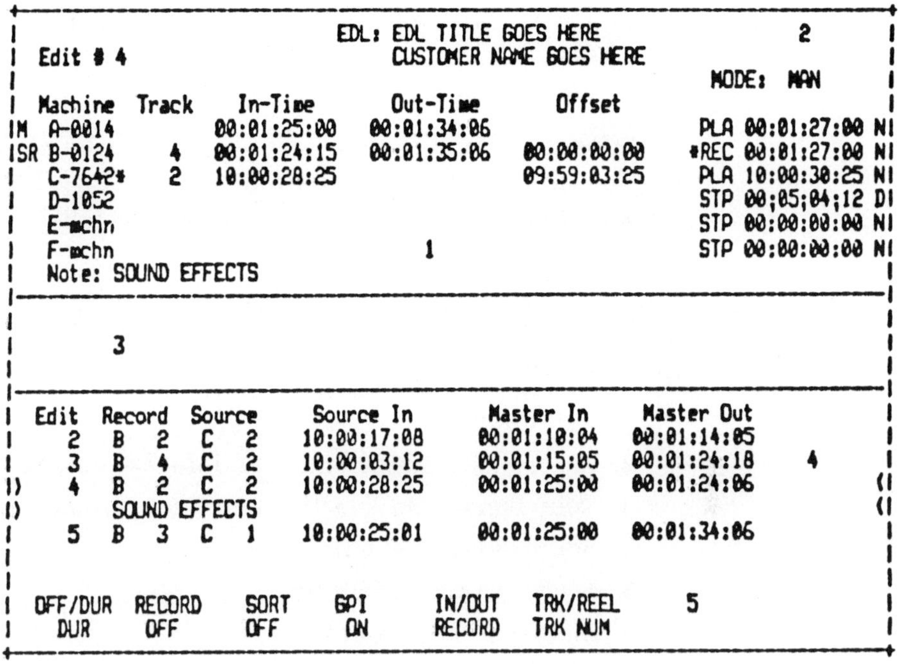

Fig. 3-23. The
edit screen of the
CASS-1. *(Courtesy
of CMX, Inc.)*

```
+---------------------------------------------------------------------------+
I                           EDL: EDL TITLE GOES HERE              2          I
I   Edit # 4                     CUSTOMER NAME GOES HERE                      I
I                                                              MODE:  MAN     I
I   Machine  Track    In-Time         Out-Time        Offset                 I
IM  A-0014           00:01:25:00    00:01:34:06                  PLA 00:01:27:00 NI
ISR B-0124      4    00:01:24:15    00:01:35:06    00:00:00:00  *REC 00:01:27:00 NI
I   C-7642*      2    10:00:28:25                   09:59:03:25   PLA 10:00:30:25 NI
I   D-1052                                                       STP 00:05:04:12 DI
I   E-mchn                                                       STP 00:00:00:00 NI
I   F-mchn                              1                        STP 00:00:00:00 NI
I   Note: SOUND EFFECTS                                                      I
I--------------------------------------------------------------------------- I
I                                                                            I
I        3                                                                   I
I                                                                            I
I--------------------------------------------------------------------------- I
I   Edit  Record  Source      Source In      Master In     Master Out        I
I     2    B   2   C   2     10:00:17:08    00:01:10:04   00:01:14:05         I
I     3    B   4   C   2     10:00:03:12    00:01:15:05   00:01:24:18     4   I
I)    4    B   2   C   2     10:00:28:25    00:01:25:00   00:01:24:06       (I
I)         SOUND EFFECTS                                                    (I
I     5    B   3   C   1     10:00:25:01    00:01:25:00   00:01:34:06         I
I                                                                            I
I   OFF/DUR   RECORD    SORT    GPI    IN/OUT   TRK/REEL       5              I
I     DUR      OFF      OFF     ON    RECORD    TRK NUM                       I
+---------------------------------------------------------------------------+
```

Detailed Edit Area—The *detailed edit area* contains the display of all information pertaining to an individual edit (referred to as the current edit). This area is divided into three sections:

1. The *machine assignment display* contains information as to master, record, slave, and current machine indicator and edit, track, and current reel number.

2. The *edit point display* lists in and out points as time code addresses for each machine involved in the current edit.

3. The *duration or slave offset display* gives offsets for the source machine and any machine slaved to the master. Alternatively, the duration of an edit can be displayed.

Machine Status Area—The *machine status area* contains indicators for the motion and addresses of the machines. This area is divided into three sections:

1. The *control mode* indicates the status (manual or automatic) of the current machine control function.

2. The *machine motion indicators* give the current transport and electronic mode for each machine.

3. The machine address indicators give the time code address for each machine and whether it is functioning with nondrop or drop frame code or (if applicable) in tach mode.

Edit Command and Status Area—The *edit command and status area* displays the edit command currently being carried out, as well as any system prompts and messages for the editing system.

Edit Summary Area—The *edit summary area* contains a display of limited information about the edits surrounding the current edit on the EDL. The following information is given for each of the edits included in the summary area:

1. Edit number
2. Record machine selection and track numbers
3. Source machine selection and track numbers
4. Source machine in time
5. Master or record in time
6. Master or record out time
7. Edit note

Mode Display Area—The *mode display area* contains the status information for the editing modes:

1. Offset or duration
2. Record mode
3. Sort type: RECORD, EDIT NUM, or OFF
4. GPI (general-purpose interface): ON or OFF
5. In and out times for master and record transport
6. Track or reel number(s)

Integral Console Synchronization System

In recent years, a number of consoles used for audio production for video have incorporated synchronization using time code into their data interfaces. One example of this type of system is the integral synchronizer and master transport selector available as an option on the SSL 6000E production console. This system enables the console to synchronize up to five machines, of which any of the first three may be designated as the systemwide master. Any or all of the remaining machines may be designated as slaves, and additional provision is made for a structure of up to 32 events per machine, or as many as 150 events per title. The entire system is operated from the built-in transport remote control unit and/or the integral SSL computer keyboard. For each production title, a "sync preset list" is created to store all sync and machine selection details for each project or session in computer memory or on diskette, along with any necessary offset values. At each session, the engineer has only to load the session program and select the desired sync preset. The computer will then automatically configure the involved machines and set the desired offsets.

Setup for Production Using Time Code

In production work where time code is utilized, the initial setup involving a combination of audio and video transports and the synchronizer requires making specific control signal connections, in addition to establishing the program signal path.

Generally, in audio production, the only required connection between the master machine and the synchronizer is that carrying the LTC playback signal (Fig. 3-24). If the synchronizer is the chase or control type, a number of interconnections must be made between the synchronizer and the slave transports. These include provision for the time code playback signal feed, for full-logic remote control over the slave transports' functions, and for a DC signal voltage routed from the synchronizer to regulate the slave transports' capstan servos (and thus transport speed).

Fig. 3-24. System interconnections for synchronous audio production.

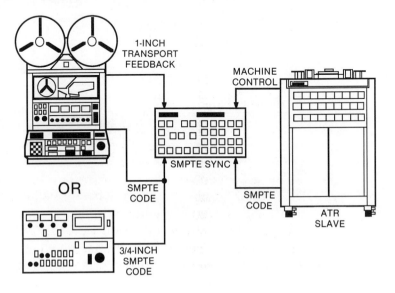

Distribution of the LTC Signal

The LTC signal may be distributed throughout the production or postproduction setup as any other audio signal is. It may be routed through audio distribution amplifiers and patched through audio switching systems via normal two-conductor shielded cables.

Since the LTC signal is biphase or symmetrical in nature, it is immune to problems of polarity as well as to inverted output amplifiers. One problem that may plague systems using time code is crosstalk; the time code signal may interfere with adjacent, low-level audio signals or recorded tape tracks.

In video production, system interconnections are often similar to those described above. However, for production involving video editing using time code and two or more video sources, there is one rule that must be followed at all times. To ensure that correct time code is locked to each video frame, the time code generator or source must be synchronized, or *gen-locked* to the video source being recorded (Fig. 3-25).

Fig. 3-25. System interconnections for synchronous video production.

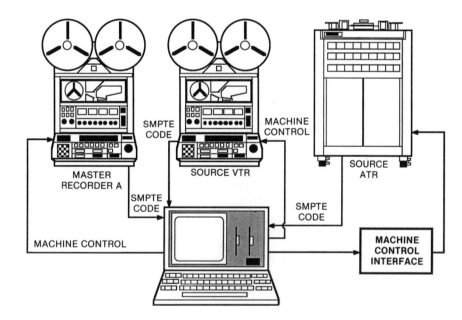

With an in-house system, where all equipment is gen-locked to a single video sync generator, all that is necessary is a separate video sync feed to the time code generator; this will lock the time code address to the composite video frame. If no house sync is available, many time code generators will accept a composite video signal as a sync source. Failure to supply the generator with a signal properly referenced to the video source will result in a nonsynchronous time code (that is, time code addresses and video frames will have no true relationship in time), a code that is useless for computer editing and many video time code applications.

Recording Levels for Time Code

The subject of track format and recording levels using time code is discussed in detail in Appendix B, but here it is worthwhile to point out one of the major problems plaguing time code production—recorded levels of the time code signals.

Currently, no industry standards exist for the levels for recording time code onto magnetic tape; however, the levels in Table 3-1 have been proven to give the best results in most cases.

Table 3-1. Optimum Recording Levels for Time Code

TAPE FORMAT	TRACK FORMAT	OPTIMUM RECORDING LEVEL
3/4-inch	Audio 1 or time code track	−5 VU to 0 VU
1-inch	Cue track or audio 3	−5 VU to −10 VU
2-inch quad	Cue track, also called auxiliary track	+3 VU to +5 VU
ATR	Edge track (highest number)	−5 VU to −10 VU

Note: If the VTR to be used is equipped with AGC (automatic gain compensation), the AGC should be overridden and the signal gain controls should be adjusted manually.

Distribution of the VITC Signal

The VITC signal, like the LTC one, may be distributed to any point in the video chain from a master generator. The VITC signal may be handled in precisely the same fashion as a camera or video line signal is. However, cable and amplifier delays must be taken into account in order to produce a properly timed signal at each inserter.

Distribution of Time Code Using ESbus

With the ever-increasing complexity of the equipment found in both audio and video postproduction suites, the demands made on a single video edit controller or audio postproduction synchronizer are such that these devices must be highly sophisticated (and are therefore expensive). In designing these devices, which maintain total operational control over multiple transports, one major requirement is often overlooked, and that is simplicity of operation. In order to move toward correcting this problem, the SMPTE and the EBU began studies intended to seek out a simplified method for distribution of time-encoded commands from a central processing unit. This method of distribution has been appropriately dubbed *ESbus* (from EBU/SMPTE bus).

In production work using multiple transports and a control-type synchronizer, all calculations of and control over tape position, speed adjustment, offset adjustment, and transport function are carried out within the synchronizer's built-in computer. In an equivalent production system using ESbus, a central control unit (CCU) communicates via a single data communications line, or bus (Fig. 3-26), with any combination of involved ATRs and VTRs, each of which is equipped with its own integral synchronizer. Each such ATR or VTR is said to be an *intelligent transport*. The basic purpose of each transport's synchronizer is twofold: to maintain a synchronous lock with the designated master transport, and to maintain total control over the transport's address location. The fact that each tape or film transport has its own dedicated synchronizer means that local control can be optimized according to the physical and electronic characteristics of the individual machines involved. Through the ESbus distribution of binary-encoded signals, which are carried over a dedicated line, the CCU is able to distribute appropriate time code addresses, whether they are auto-locate (go-to) or synchronized (designated master) addresses, and to provide remote control of designated transports or event-related devices. The CCU acts as an information distribution device and does not provide actual locations, so related calculations, device interconnections, and user operation are all greatly simplified.

One such system in current usage is the TLS 4000 from Studer/Revox of America, Inc. (Fig. 3-27), which embodies a "building block" or expandable approach to a synchronization system. The most basic form of the TLS 4000 utilizes the "Black Box" synchronizer, which functions as a simple parallel control interface between two transports (Fig. 3-28). With the addition of the system's local control unit, or LCU, a greater degree of control may be exercised over one or more transports in a limited control-type sense (Fig. 3-29). With the incorporation of the SC 4000 system controller (a CCU), up to 16 transport units may be operated (through individual "Black Box" interfaces) under tandem ESbus control (Fig. 3-30).

Fig. 3-26. Use of ESbus for system interconnections in audio-for-video production.

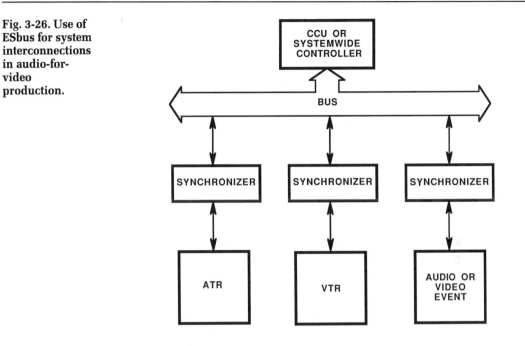

Although the ESbus technology is still in the formation stages, the potential for such systems is readily appreciated. For example, with ESbus, a CCU may be designed as part of the audio production console, directly tied into its communications structure, along with automation and signal-routing functions. Thus, with a future system that incorporates a compatible ESbus-related data structure, every aspect of an audio-for-video postproduction mix may be stored within the internal memory of the console.

Fig. 3-27. Studer "Black Box" controller/synchronizer and TLS 4000 synchronizer. *(Courtesy of Studer/Revox of America, Inc.)*

**Fig. 3-28.
"Black Box"
synchronizer
with simple par-
allel control.**
*(Courtesy of
Studer/Revox of
America, Inc.)*

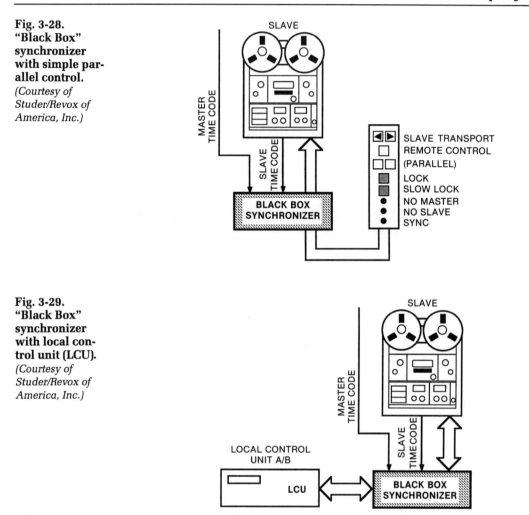

**Fig. 3-29.
"Black Box"
synchronizer
with local con-
trol unit (LCU).**
*(Courtesy of
Studer/Revox of
America, Inc.)*

The VTR as Slave in Postproduction Using Time Code

When a VTR (or VCR) is used in a slave capacity within a synchronization system, constraints are imposed as a result of the way in which a
VTR operates.

Unlike an ATR, a VTR has two control circuits for servo speed, one
to control tape speed and one to control the speed of rotation of the
video record/reproduce heads. The tape and the heads must always operate in precise synchronization in order for the video signals to be re-

**Fig. 3-30.
Complex syn-
chronization
system with cen-
tral controller
and various
slave machines.**
*(Courtesy of
Studer/Revox of
America, Inc.)*

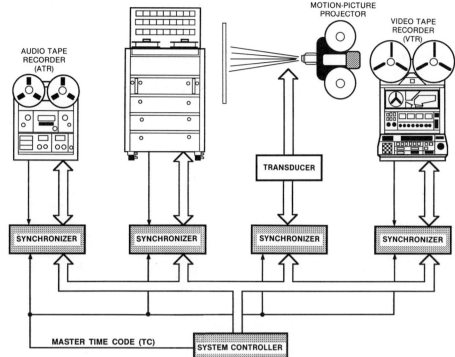

corded or reproduced correctly (although a video signal is actually an analog signal, it is handled in a discontinuous fashion, as lines, fields, frames, etc., and is precisely related to the video sync signal). Therefore, it is not possible for an ATR control track to be used to provide continuous synchronization control over a VTR. To do so would prevent the tape from running in sync with the rotating heads, and thus the video signal would not be recorded or reproduced properly.

Because of these facts, a synchronizer that is wired for control over a VTR operating as a slave must function somewhat differently than a synchronizer controlling an ATR. One technique that may be used is to allow the synchronizer to control the VTR's tape speed until the time code addresses of master and slave have been synchronized, after which this control is released, permitting the VTR's tape transport servo to lock the tape to the video sync using its own circuitry. In order for the slave and master recorders to continue running in sync after control of the slave's capstan speed has been released, all VTRs within the system must be fed an appropriate video sync signal originating from the same source.

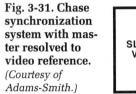

**Fig. 3-31. Chase
synchronization
system with mas-
ter resolved to
video reference.**
*(Courtesy of
Adams-Smith.)*

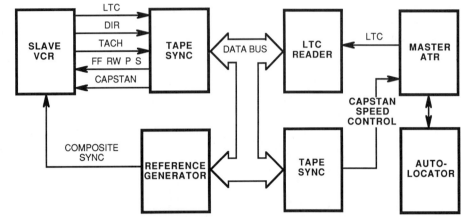

For audio-for-video editing, it is sometimes convenient to use a multitrack ATR as the master and a VTR as the slave, thereby assigning system transport control to the ATR's auto-locator or to an ATR control synchronization system. Because of the way in which a VTR must be operated when it is in a slave capacity, it is not sufficient simply to interchange or repatch recorder controls. In order to keep the slave's videotape and the master's audiotape running in sync after control of the slave's capstan speed has been released, it is necessary that the ATR be synchronized to the same reference signal as the VTR, at nominal play speed. This technique of running an ATR in sync with a reference signal is sometimes called *resolving* and requires the use of two time code synchronizers and a reference generator (Fig. 3-31), such as the Adams-Smith 2600 RG.

Film Production and Postproduction Using Time Code

Production applications for editing film using time code are still largely experimental. The foremost problems are encountered in dealing with the different frame rates for film and videotape. The frame rate for film is 24 fr/s while videotape runs at approximately 29.97 fr/s (SMPTE) or 25 fr/s (EBU). In order to make each second of film match up precisely with each second of videotape using time code synchronization, a different time code rate would have to be introduced (a drop frame code for film). This would prove difficult as extensive frame-matching calculations would be necessary between the two formats. One proposal for time code identification of film is the adoption of a code that does not allow for the inclusion of frames but still includes hours, minutes, and seconds for both film and sound.

Provision for synchronization with film is being incorporated into some of the newer control and EDL-based synchronizers through a facility for breaking down the frame timing increments into smaller subdivisions. For example, the CMX CASS-1 offers a *frame-bump feature*, which allows a user to move one transport's tape in reference to another's in increments of one frame or in user-programmable increments of hundredths of a frame. Thus, to sync a 35-mm magnetic film track to videotape requires that the frame-bump value be set to 25/100 of a frame, which equals one sprocket hole.

The ability to synchronize sprocketed magnetic film to videotape or audiotape may be added to the Adams-Smith System 2600 by incorporation of a recently developed option called PCB. The PCB device outputs TTL-level, biphase (quadrature-phased) square waves, with a nominal frequency of 240 Hz, 250 Hz, or 300 Hz (or other required frequency) at normal transport play speeds. With this option fitted, the film transport can be cued and synchronized with frame accuracy, using LTC recorded onto one track of the film, just as it is with an ATR. Alternatively, the film transport may be operated without the use of time code, istead using the biphase square wave as a reference pulse during synchronous running (as pilot tone signals are used with phase-locked ATRs). To accommodate system configurations in which the film transport is controlled by other equipment, the PCB option can accept biphase signals from the transport and decode them into a tach pulse and direction signals that are used in updating time code readings.

Regardless of which production method is used for the creation of a motion picture (film-video, video-film, film-film), time code is still the most accurate and cost-effective means for producing film sound. This is true in building dialogue, music, and effects tracks (SFX), and even in the synchronizing and laydown of all recorded material onto a multitrack tape machine for the production of a final mix.

4 *Audio Production for Video*

The first stage involved in the creation of quality audio for video is the *production phase*. This is when much of the original source material for later postproduction may be recorded.

The production of video takes on a multitude of forms, and the choice of one often depends on the environmental conditions under which the video shoot is to take place. The following are a few examples of production methods that have been developed for specific conditions:

- ENG (electronic news gathering)
- On-location EFP (electronic field production)
- On-location use of a van and mobile facilities
- Utilization of a video production facility
- Shooting video to prerecorded audio tracks

Conditions often dictate the type of production method required in order to follow the program or script material. For example, it may be far less expensive and time-consuming to shoot an outdoor crowd scene using EFP at a location site than to build the necessary set within a production stage. The wide variety of production situations that are encountered means that a large repertoire of recording skills are required of the audio-for-video professional, who must sometimes specialize in a particular production method.

This chapter discusses many of the production methods and devices used in the production of professional-level audio for video. Also included is a detailed look at one of the major tools available to the audio production specialist—the microphone.

The Recording Chain

Before the thorough examination of audio production for the video medium, there is one concept that should be examined at this point. This is the concept of the *recording chain*. The recording chain is governed by a rule that is valid for all electronic media:

The overall quality of a signal is no better than the quality of each of the individual components that combine to make up that signal.

Central to the idea of the recording chain is the concept of the *transducer*. A transducer is any device that is capable of changing one form of energy into another form of energy. For example, an violin is a transducer (Fig. 4-1); the vibrations caused by the bowing or plucking of the strings are amplified through a body of wood, which in turn converts these vibrations into corresponding pressure waves in air, which we hear as sound. A microphone (mic) is yet another example of a transducer. Sound waves act on the microphone's diaphragm and are converted into a series of corresponding electrical voltages that may be amplified or recorded.

Fig. 4-1. The violin and microphone as transducers.

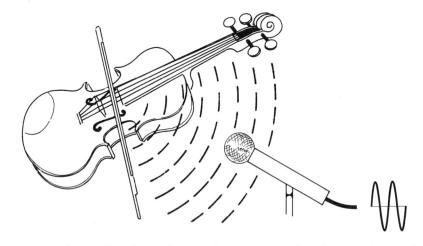

Within the audio chain for a video project, the electrical impulses generated by a microphone are often fed to a portable ATR or to the audio track of a VTR. There the electrical energy is converted into magnetic force fields, called flux, and applied to magnetic tape. The tape stores these magnetic fields in the order and orientation in which they are presented to it, so that the electrical impulses that generated them can be recreated during playback at any later time (Fig. 4-2). The recreated electrical impulses may be dubbed over (with the associated analog generation losses) to a multitrack ATR for postproduction sweetening and may then be layed back to the edited master (EM). Once in its final form, the recorded material may be broadcast or distributed in a reproduced format. In either case, it is reconverted back into electrical energy and finally into electronic mechanical energy at the loudspeaker, which in turn recreates the original variations in air pressure that were sensed by the microphone. As can be seen, transducers are to be found at practically every point within the audio chain.

A transducer is often found to be a weak link within the recording chain. Given the present technology, the transduction process cannot be accomplished perfectly. Noise, distortion, and often coloration of a sound may be introduced to a greater or lesser degree, although these defects can be minimized with care. Differences in design are another factor; even the slightest difference in design between one microphone and another can possibly cause major sound differences. This factor and the complexity of sound, acoustics, and recording techniques combine to make sound recording a very subjective endeavor.

With digital recording, an audio system may have the major advantage of allowing for a significant reduction in the number of transducers found within the recording chain (Fig. 4-3). In a digital recording chain, the acoustic waveforms picked up by the mic and converted into electrical impulses are often passed through a series of amplification and processing steps within the recording console. Once these processed electrical impulses have been delivered to the digital ATR, they are converted into digital form by an analog-to-digital (A/D) converter. At this point, the electrical signals are converted into digital numeric values, representing corresponding voltage levels. These signals will thereafter remain in a digitally encoded form, through any intermediate steps, until they finally pass through a digital-to-analog (D/A) converter (such as in a digital satellite linkup or a compact disc player). Thus, if a digital recording is listened to at home, the minimization of transducers in the recording chain, along with other advantages of digital recording, will result in a notable sonic clarity, equal to that of the original studio master.

Audio Recording Systems in Video Production

The first stage encountered within the recording chain for production of quality audio for video is often the recording of dialogue, back-

Fig. 4-2. The conventional analog recording chain.
(Courtesy of Sony Corporation of America, Inc.)

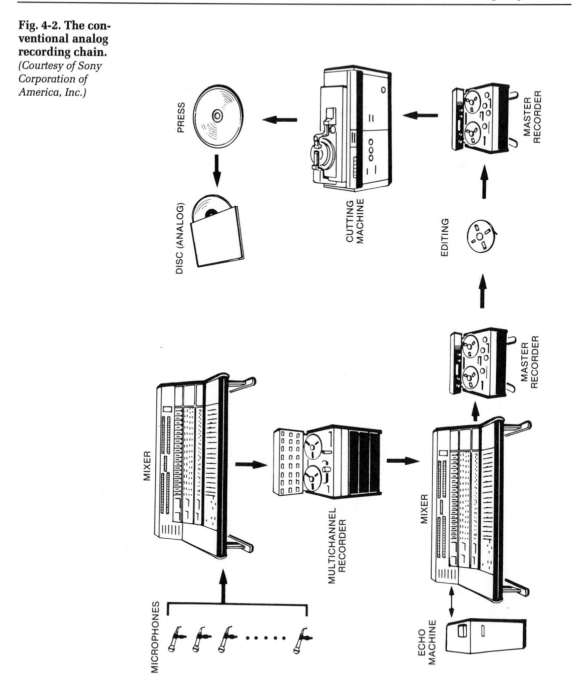

Fig. 4-3. The all-digital recording chain. *(Courtesy of Sony Corporation of America, Inc.)*

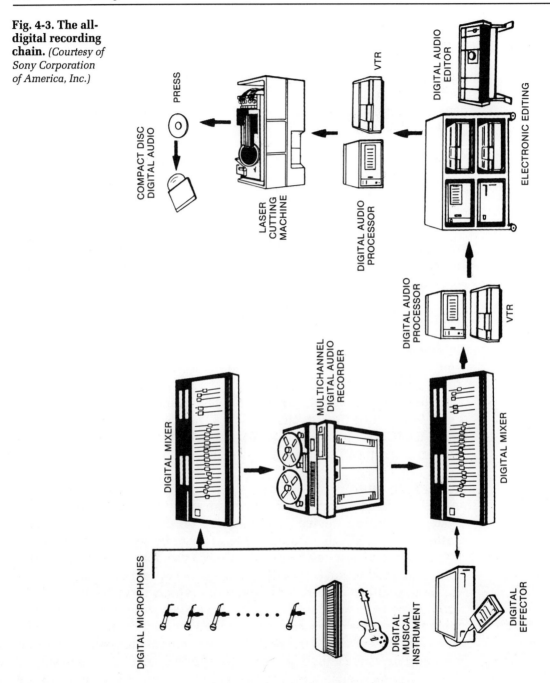

ground presence, and sound effects at the live video shoot. It is extremely important that care be taken to obtain a quality, first-generation recorded signal. Although it is possible to correct for a less than perfect quality of the recorded audio field master, there is no true substitute for the realism and involvement that are the result of the accurate capturing of sounds present at the original shoot. In order for the fullest advantage to be taken of the modern audio postproduction process, it is necessary that the original recorded sound tracks be clean in basic character and that they show a degree of isolation from background noise or presence sounds (which may be recorded onto a separate track). This will allow for the greatest degree of control and flexibility during the later stages of production.

Basically, two production methods exist for the recording of synchronous sound onto an audio field master at the time of the original video shoot: single-system sound recording and double-system sound recording.

Single-System Sound Recording

In *single-system sound recording*, the medium used to record the visual image is also employed for the recording of sound. In the past, film cameras were fitted with audio recording capabilities for production; however, the vast majority of single-system sound recording presently being done utilizes the VTR. This method of recording tends to lend itself best to smaller single-camera (single-cam) production, within an on-location setting.

The Porta-cam

The *porta-cam* (or portable camera) is an excellent example of a single-system recording device in day-to-day operation. *Electronic news gathering*, or *ENG*, the product of which can be seen daily on almost any local television news program, makes use of the porta-cam in on-the-scene production (Fig. 4-4). The device's popularity is due to its extreme portability and ease of operation, which serve to reduce both production time and field crew. *Electronic field production*, or *EFP*, in which all or some of the scenes for a video production are shot on location, also makes use of the porta-cam single-system method of audio production. This method is especially applicable in the case of single-cam or limited-budget productions. Portable systems such as the Sony Betacam are self-contained, over-the-shoulder units incorporating VTR, ATR, and camera that provide a high degree of versatility in ENG and EFP.

Fig. 4-4. Photographer/ editor Alice Webb, of Metromedia's KRIV-TV, uses one of the station's Betacam® camera/VTR systems to get a news story on the streets of Boston. *(Courtesy of Sony Corporation of America.)*

One of the simplest techniques for recording audio using the single-system method is *in-camera recording* of the original sound, that is, recording the on-camera sound using a microphone that is housed directly on the body of the porta-cam itself (Fig. 4-5). In professional systems, this mic is generally of the directional type, often a highly di-

Fig. 4-5. A porta-cam with built-in microphone. *(Courtesy of Swintek/ Magnavox.)*

rectional shotgun mic. Such a microphone allows for on-camera sound to be recorded while the majority of off-camera sound is rejected.

The External Microphone

Another single-system recording technique makes use of an *external microphone*, which may be either a hand-held or lavalier microphone. This method serves to satisfy the basic requirements of clean pickup and rejection of unwanted background sounds. These are fulfilled simply because the mic is in such close proximity to the on-camera event (generally on-camera talent).

The Wireless Microphone

A third single-system method of recording is one utilizing a *wireless microphone*. Such a method of sound pickup allows for the greatest degree of mobility and freedom for the sound source with respect to the video camera and VTR, while maintaining a clean audio signal. A wireless microphone system (Fig. 4-6) has three components: a pickup unit, a receiver, and an antenna. The pickup unit within such a system is itself made up of two components: the microphone and the frequency-modulated transmission device. This combination may take any of several forms, for example, hand-held microphone and body transmitter, lavalier microphone and body transmitter, or integrated microphone and transmitter. The receiver component picks up the VHF or UHF frequency-modulated signal that is sent out by the transmitter. Receivers may be rack-mounted base stations or smaller portable devices for use with ENG or EFP. The antenna system may be designed as either integral or external to the overall wireless system. Two more examples of wireless microphone systems are shown in Figs. 4-7 and 4-8.

Double-System Sound Recording

For single-cam productions, where portability and ease of operation are primary production concerns, single-system sound recording is generally sufficient. However, when multicam (multiple-camera) production is called for in an EFP shoot or when a higher degree of audio quality is required to effectively reproduce the desired on-camera sound, it is often necessary to employ the techniques of *double-system sound recording.*

As the name implies, with double-system sound recording the audio signal is recorded onto a separate medium from that on which the

Fig. 4-6. The Sony VHF-synthesized wireless microphone system. *(Courtesy of Sony Corporation of America, Inc.)*

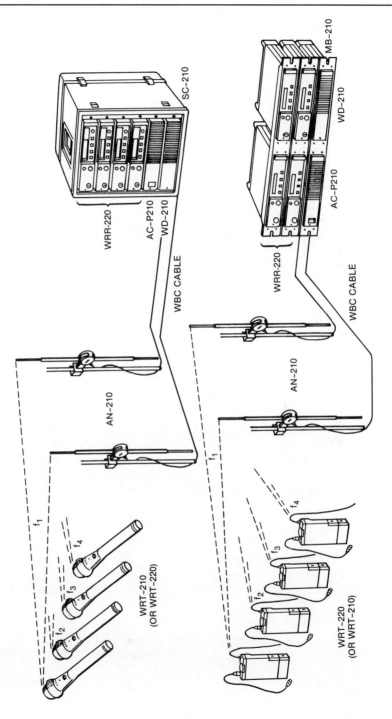

Fig. 4-7. Swintek
wireless micro-
phone system—
2L-RSFD with
MU50A and
SM-57 hand-
held. *(Courtesy of
Swintek.)*

Fig. 4-8. Cetec
Vega PRO-1
wireless micro-
phone system.
*(Courtesy of Cetec
Vega.)*

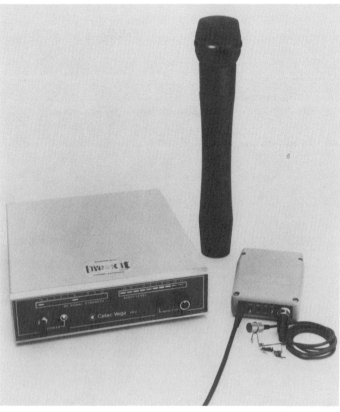

video signal is recorded. In this method of production, the audio signal is recorded directly onto an ATR, and the video signal is recorded onto one or more VTRs. This approach is somewhat more complicated than the single-system method and may require a larger and more specialized production crew. However, this type of system allows the audio professional a greater degree of control and versatility. This results from the sound system's independence from the video recording chain, which can mean a higher degree of recorded audio quality.

Synchronization and Double-System Recording

In order for the video and audio portions of a program to display a precise time relationship with respect to each other using double-system sound recording, some form of precision synchronization must be used.

The Pilotone method for achieving synchronization is still used to some extent worldwide in the production of audio for film. However, in current practice in audio production for both video and film, the most accurate and commonly encountered method for achieving synchronization is through the use of time code. The incorporation of time code in double-system production is accomplished by setting all of the associated transports to the same nonslip time code address. The use of such a synchronizing address allows for the simultaneous and independent production of the audio and visual aspects for postproduction processing. Changes can be made during postproduction to the program content of each of these aspects, and finally they can be rejoined in synchronization as the final product (most often taking the form of an edited master videotape).

House-Synchronization—In the medium of video, it is of primary importance that the frame stability of the visual image be perfectly maintained from one piece of video equipment to another within the production facility. Therefore, it is absolutely necessary that the video picture timing be accurately controlled throughout the system. This is most often accomplished through the use of a single house video sync generator, which is said to provide *house-synchronization*, or *house-sync*. This leads directly to one of the most important and widely applicable rules for the production of video and audio for video (Fig. 4-9):

> The time code generator used to provide housewide time code must be synchronized (gen-locked) to the video source being recorded.

Fig. 4-9. In an in-house video production facility, the time code generator must be gen-locked to the house video sync generator.

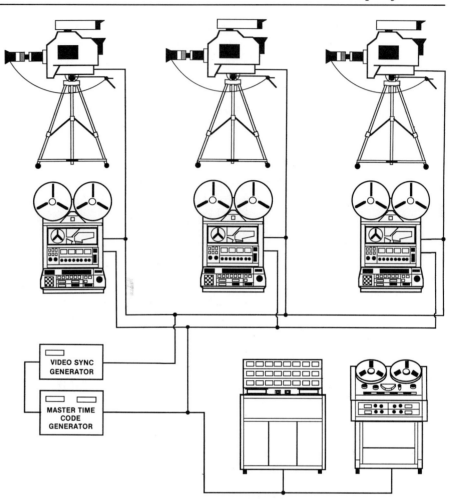

This is of particular importance in video production when on-line editing or editing based on time code addresses is to be done during postproduction.

Chapter 3 described how one 80-bit time code word is used to mark off one video frame, which is made up of two specific fields for color reproduction. In the video editing process, when time code is used, it is absolutely necessary that the starting point of the time code address fall exactly at the starting point of the vertical signal (as denoted by the vertical interval). When these two points are in synchronization, the time code signal and video signal are said to be gen-locked. Failure to gen-lock these signals will result in the recording of nonsynchronous time code, which is generally useless for computer editing and most other applications based on time code. For example, within postproduction, a video editor may want to make an A/B switch from one VTR to another at a point in time when their video signals are at differing points of the frame cycle, almost assuredly resulting in a picture breakup.

Within a house system, where all associated video equipment is gen-locked to a common video sync generator, all that is required for gen-locking the time code is to run an additional feed to the master time code generator from the video sync source.

If no house-sync is available, the majority of time code generators currently on the market are designed to accept any sync signal (such as a composite video signal) directly from the video source whose output is being recorded (Fig. 4-10).

Fig. 4-10. Most modern time code generators will accept a composite video signal as a video sync source.

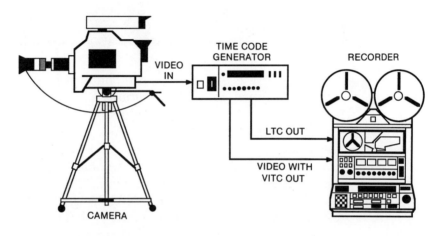

The Portable Professional ATR

Given the diversity of working environments and approaches to production encountered in the field, two of the greatest assets a piece of professional audio equipment can claim are versatility and portability. For many decades, the portable professional ATR proved itself in the production of high-quality on-location and in-studio audio for the motion picture industry. This same machine continues to make itself indispensable in the double-system recording of audio for video during EFP.

The Nagra IV-S
The Nagra line of portable ATRs has long been proven to be the workhorses of the audio-for-video and audio-for-film production environments. One of the most popular of these Swiss-made machines is the

**Fig. 4-11. The
Nagra IV-S
portable ATR.**
*(Courtesy of Nagra
Magnetic Re-
corders, Inc.)*

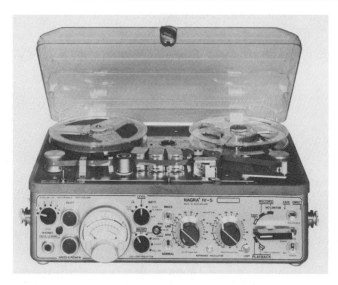

Nagra IV-S (Fig. 4-11). This self-contained stereophonic tape deck comes equipped with many features and has additional options that make it well-adapted to the production of audio for the visual media. In its standard configuration, the IV-S operates with standard 1/4-inch mastering tape at transport speeds of 15 ips, 7-1/2 ips, and 3-3/4 ips. The recording format consists of two audio tracks and one sync track for external Pilot-based synchronization. Tape equalization is user-selectable to NAB, DIN (CCIR), or NAGRAMASTER. Various internal accessories allow for many standard appointments, such as balanced microphone line inputs, phantom powering (at various standard voltage levels), microphone line equalization, and input signal limiting. Internal accessories such as Pilot-signal metering and a crystal-controlled generator for the Pilot signal are included. External accessories include a Pilotone synchronizer (to phase-lock the playback speed to an internal or external reference frequency), external system option (noise reduction, compressor, etc.), and VHF transmitter/receiver for camera-to-recorder slating by remote radio.

The Nagra IV-S with Time Code—Through the application of technology developed for the Nagra/Ampex VPR-5 video tape recorder, Nagra has been able to integrate a complete SMPTE/EBU time code generator and record/playback system into its IV-S ATR, thereby creating the Nagra IV-S TC (Fig. 4-12). A waferlike readout-and-control panel retracts under the front panel of the ATR for ease of access. On this panel, all time code addresses and status commands may be easily read from a liquid crystal display (LCD), and entries of addresses and commands may be made on the touch-panel controls provided. The internal time code generator may be gen-locked to house video sync or may be slaved to a master time code generator through continuous jam-sync. User data may be entered into the time code word, and the frame number may be preset to zero or a particular time of day to provide for house-sync. This ATR may be used similarly to a wireless microphone system in double-system recording by jam-syncing it to the master generator at the beginning of the shooting period and operating it independently of the video system thereafter.

Fig. 4-12. Nagra IV-S equipped with time code. *(Courtesy of Nagra Magnetic Recorders, Inc.)*

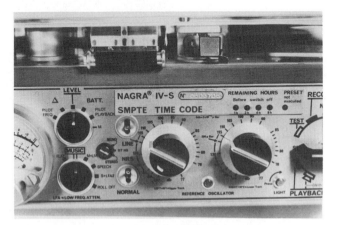

The Nagra T Audio

A unique addition to the Nagra family of portable ATRs is the T Audio (Fig. 4-13), which was developed from the start as a postproduction machine based on time code. An optional internal synchronizer may be fitted to ensure the slaving of the T Audio to any external time code source in accordance with the SMPTE/EBU standard. The integration of one or more T Audio machines into an audio-for-video or audio-for-film postproduction and editing system may be facilitated by the optional RS 422 ESbus interface, which conforms to the universal studio bus standard. Another available interface allows the unit to communicate with a computer system.

The T Audio incorporates a removable head block, which includes a separate time code erase head (between the audio erase head and tape tension head) and a time code record/play head (located in the middle of the block just before the audio record head). At the rear of the machine, a supplementary panel carries all of the connectors associated with time code, that is, time code signal I/O, internal synchronizer, and ESbus serial ports. On the front panel, a lemo-type connector (the same as that used on the Nagra IV-S TC, the Ampex/Nagra VPR-5, and the

**Fig. 4-13. Nagra
T Audio portable
ATR.** *(Courtesy of
Nagra Magnetic Re-
corders, Inc.)*

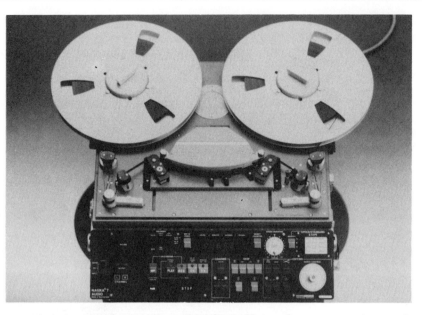

Nagra mini time code generator/reader) allows for easy setting and re-setting of the internal time code generator.

Four frame rates are available when using the T Audio: 24 fr/s (film, SMPTE), 25 fr/s (film/video, EBU), 29.97 fr/s (color video, SMPTE), and 30 fr/s (monochrome video, SMPTE). Selection is accomplished using an internal switch located at the front of the time code circuitry. The 29.97-fr/s rate may be generated either with or without drop frame code.

When the main power is disconnected, the unit's internal clock and RAM (random access memory) continue to be powered by batteries, ensuring the contents' autonomy for more than one year.

The internal time code generator may not be utilized, or it may be set using either the T Audio's control unit or an external time code source. In the assembly mode, the device will synchronize to time code read off tape, but will switch to the internal generator as soon as recording begins (jam-sync). Since the time code and audio signals are read by two different heads, the microprocessor control unit must cal-culate a time increment to be added in order to compensate for the dis-tance between the heads. The code is incremented upward, depending on the tape speed, so that the time code addresses are correctly corre-lated to the tape.

The Time Code Reader of the T Audio—The circuitry of the time code reader of the T Audio is in fact a double playback system, allowing the microprocessor to have simultaneous access to the time code signal read from the tape and an external TC source (such as another ATR, a VTR, an editing system, or a master clock). The playback preamplifier is fitted with an AGC (automatic gain control) circuit, a tracking filter, and a sophisticated, self-adapting, threshold comparator that allows for reliable time code playback at 0.25 to 70 times the nominal transport speed. The decoding systems for the playback and external signal also have a range of 0.25 to 70 times the nominal transport speed.

At less than 0.25 times the nominal transport speed, the displayed time code value can be incremented by the pulses coming from the tape counter tachometer wheel. The time code signal is available at the output connectors in three forms:

1. As read by the time code, that is, in analog form, preamplified, filtered, and regulated.
2. As read by the time code head after the comparator, that is, in logic form, without error correction and in advance of the audio signal.
3. As supplied by the microprocessor, that is, compensated, regenerated, and correlated with the audio signal, with playback errors (dropout, etc.) automatically compensated for and possible components of wow and flutter filtered out (note that the signal, as described here, is available only within a range of $\pm 12\%$ of nominal speed).

The Pilot Signal Reader, Demodulator, and Synchronizer of the T Audio—In order to ensure the compatibility of all tapes recorded using the various Nagra Pilotone systems, optional features enable the Nagra T Audio to operate as a universal transferring machine, which may be synchronized to time code or FM Pilot, as well as to the Neopilot signal. An additional circuit may be fitted to the T Audio that allows for integral playback and demodulation of the Pilot signal and also requires that an additional Neopilot sync head be added in tandem. Switching between the employed heads may be accomplished using a switch located on the FM Pilot board.

In its version including time code and synchronizer, the T Audio offers accurate synchronization at nominal play speed, in shuttle mode and in fast wind mode, utilizing features usually found only in larger studio ATRs:

- The twin capstan, closed-loop tape transport has a low inertia, which ensures an accurate monitoring of tape speed and therefore significantly reduces the rollback time compared to that with conventional ATRs.

- The "Listen TC" mode causes the pinch wheel assembly to move forward, bringing the tape into contact with the time code playback head for synchronous chase purposes.
- The manual servo-control wheel allows optimization of the offset value to 1/80 frame (1 bit); the offset value entered on the control unit is programmable from −23:59:59:29:79 to + 23:59:59:29:79 (HH:MM:SS:FF:BB) for the 30-ips format.

Portable Professional Mixing Consoles

As the requirements for improved production within sound stages and on location have increased in recent years, so has the need for compact professional audio mixing consoles grown. The inception of transistorized mixers for field production in video broadcasting and film sound placed certain limitations on EFP audio production. Such devices generally offer only a limited number of inputs and signal control options (for example, two microphones, one line, RIAA, and rolloff filter). However, with the growth and improvement in quality of audio-for-video techniques, many of the features found on much larger production consoles, such as increased equalization and signal routing capabilities, have recently been made available on more compact configurations.

EELA Audio S-191
The EELA Audio S-191 mixing console (Fig. 4-14) is a 19-inch, rack-mountable unit offering six microphone/line (mic/line) inputs and a two-channel stereo output. Phantom power, balanced mic/line inputs, VU or European PPM metering, and P&G (Penny and Giles) faders are supplied as standard equipment, along with three-band equalization and two auxiliary sends for each channel. Limiters, or compressors, are also supplied as a standard element within the main output buses. The EELA Audio S-191 is part of the S-100 series, which includes frame sizes of up to 16 channels.

Calrec Mini Mixer
Another portable mixing console, whose design is similar to that of the EELA Audio S-191, is the Calrec Mini Mixer (Fig. 4-15). This system for professional recording and broadcast use is available in table-top, drop-in, flight case, or 19-inch (with only 8 input channels) versions, all rack-mounted. It is designed to accommodate up to 16 input chan-

Fig. 4-14. EELA Audio S-191 portable mixing console. *(Courtesy of ESL, Inc.)*

nels, each with four auxiliary prefade and postfade sends, high-frequency and low-frequency filters, and three-band equalization. Billed as a dual-channel production device, the Mini Mixer is equipped with a unique two-group, or stereo A/stereo B, output mix capability, allowing for the increased versatility of subgroup or separate feeds. A master fader is also provided for doing a separate monaural mix at the output.

Fig. 4-15. Calrec Mini Mixer. *(Courtesy of Audio + Design/ Calrec, Inc.)*

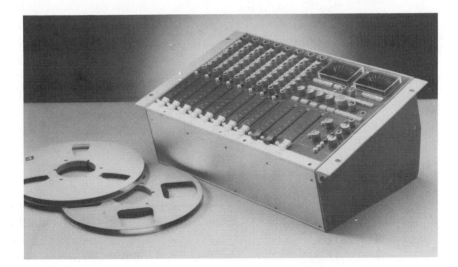

Multitrack Audio in Double-System Production

Since its inception more than two decades ago, the multitrack ATR has been the standard means of production for the music recording industry. Only recently, however, has the video industry begun to take advantage of the flexibility to be gained in production and postproduction processes through incorporation of multitrack audio techniques.

There are many possibilities for using the multitrack ATR within the postproduction building of a modern video soundtrack. However, there is another possible application for this machine that should be given consideration—utilization of the multitrack ATR in the audio production phase. The multitrack ATR may prove a beneficial tool as the demand for high-quality audio in EFP increases. In EFP, the use of a multitrack ATR would add to the amount of equipment required in the field. However, it would also provide a greater degree of flexibility with respect to the final mix of the sound track.

In conventional studio production of a soundtrack for video, the field audio is recorded onto one or two audio tracks on the videotape. In such a situation, if two or more mics are used in a dialogue scene, the sound recordist may be faced with the need for creating a *sub-mix* (where two or more audio signals are mixed onto one signal path, or track). Due to production limitations, the mixed signals may lack the required balance or quality. In that case, the scene may need to be reshot, corrected through gain riding of a fader during the mix, or corrected through replacement of the dialogue in the postproduction phase. The advantages of multitrack recording in video production lie in the elimination of the need for a sub-mix, since each audio element can be recorded onto a separate track. A similar benefit exists for the production of multitrack audio for stereo in that control can be exerted over each audio element separately for added flexibility and realistic stereo imaging during the mixdown phase.

Shooting Video to Prerecorded Audio

In certain production situations, it is often convenient and cost-effective to shoot video program footage to prerecorded audio tracks (Fig. 4-16). For example, this is usually the case in the shooting of a music video, where the recorded track is a preproduced studio soundtrack.

Fig. 4-16. Simplified diagram of the setup for a video shoot with synchronous playback.

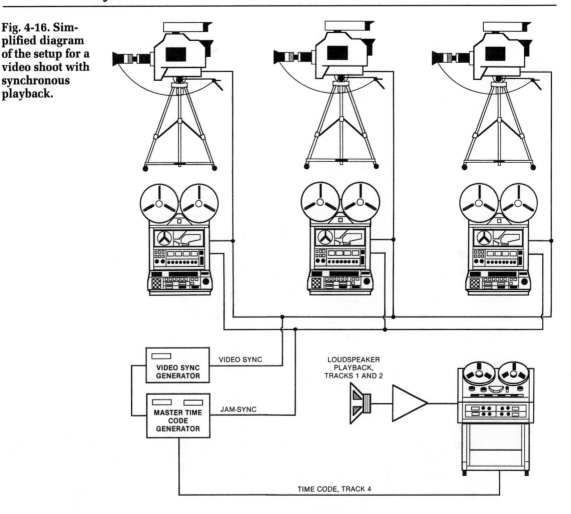

VIDEO SYNC GENERATOR

VIDEO SYNC

MASTER TIME CODE GENERATOR

JAM-SYNC

LOUDSPEAKER PLAYBACK, TRACKS 1 AND 2

TIME CODE, TRACK 4

When video is to be shot to prerecorded audio, it is first necessary to obtain a time-encoded copy of the audio master tape. The format of this master tape is generally accepted to be the 4-channel, 1/2-inch format, which is configured as follows:

Track 1: Mix right

Track 2: Mix left

Track 3: Guard band

Track 4: Time code

This 4-channel time-encoded audio master or an equivalent dubbed copy (a wise precaution) serves as the final layback production master.

If a master does not exist in the recommended format (or an acceptable center-code format), a copy must be made. This may be accomplished by either of two methods. If the multitrack master tape contains time code and was mixed utilizing the automated mixdown process on a suitably equipped console, then a mix may be performed directly from it to the 4-channel format. If this is not the case, then a direct dub of the master tape may be made to the 4-channel format while stripping master time code onto audio track 4. It is important to keep in mind that this code will serve as the master time code for the subsequent project and therefore should be set to start at time 00:00:00:00, with the music track to begin following a one-minute pre-roll (that is, at 00:01:00:00), which will provide adequate sync-up time.

Once the 4-channel master has been prepared, a reference dub to be used for playback during EFP must be also made. This dub is made in the same format as that of the ATR to be used at the EFP site, for example, 1/2-inch, 4-channel, 1/4-inch, 2-channel (with center track time code), or 1/4-inch, 1/2-track (with channel one containing a monaural track mix and channel two containing time code). It is important when dubbing time code from one machine to another (regardless of formats) to regenerate the signal at the synchronizer through the use of the jam-sync function. This is done in order to prevent the deterioration of the time code signal at later stages of production (an editor or layback engineer's nightmare).

Once the playback dub has been prepared, the next step is the actual video shoot, whether on stage or by EFP. In a production in which video is shot to audio, the playback dub serves as the master in providing both playback monitoring over loudspeakers and reference time code. During playback, the regenerated time code track is used as the production house-sync, thereby locking all field ATR or VTR masters to the time code that was present on the original 4-channel audio master tape. The composite video signal should also be synchronized (gen-locked) with the time code signal by way of appropriate interconnections between the time code generator (providing the jam-sync signal) and the video sync generator or gen-lock inputs.

Once the audio and video signals are in sync, the shoot may begin. The system may be brought into sync at any point during the program (such as 00:02:34:47), since a clean time code signal and video sync are the only requirements for the video edit and layback process.

After the shoot is completed, the video field masters may be dubbed to 3/4-inch video cassettes for the edit decision phase, and then a final video edited master (EM) is created. Finally, the audio portion is resynced from the original high-quality 1/2-inch, 4-channel master tape (containing the originating time code track) to the edited video master in the layback phase.

Microphones and Their Characteristics

In audio production for either video or music recording, the *microphone (or mic)* is very often the first device in the recording chain. Earlier in this chapter the point was made that the microphone is a transducer; its function is to change one form of energy (acoustical) into another form of energy (electrical). As a transducer, the microphone (and therefore the quality of its pickup) is subject to both external variables (the acoustic environment and its placement within this environment) and internal variables (such as the electromechanical design of the transducer itself). All of these interrelated elements work together to affect the sound quality of a microphone, and each type or design of this very subjective device has its own particular sound characteristics and applications.

Three types of microphones are currently used professionally for audio pickup and recording: the dynamic microphone, the ribbon microphone, and the condenser microphone.

The Dynamic Microphone

The *dynamic, or moving coil, microphone* is a pressure-operated device consisting of a finely wrapped coil of wire that is attached to a delicate movable diaphragm (Fig. 4-17). This attached coil is suspended within a permanently charged magnetic field. When a sound wave hits the diaphragm, the diaphragm and the connected coil are displaced in proportion to the intensity of the oncoming wavefront. By this displacement, the coil is made to cut across the many fixed lines of magnetic flux supplied by the permanent magnet, thereby inducing an electrical current into the wires of the coil. This becomes the analogous electrical output of the microphone.

Fig. 4-17. The dynamic microphone.

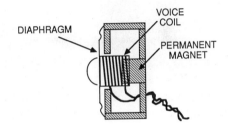

One basic rule concerning the generation of electricity is that whenever a conductive metal (such as that making up a microphone coil) is moved across the flux lines of a magnet, a flow of current (an electrical signal of a specific level and polarity) is induced in that metal. The level and polarity of this electrical current is proportional to the number and direction of the lines of magnetic flux. With respect to the diaphragm of a dynamic microphone, the amplitude and frequency of the pickup are directly related to the speed and direction in which the coil is made to cut these lines of magnetic flux.

The Ribbon Microphone

The *ribbon microphone* (Fig. 4-18) is a pressure-gradient microphone. It utilizes a thin metal ribbon that is suspended between the two poles of a magnet. When this ribbon is displaced by the pressure variations of sound waves, it cuts through the lines of flux generated by the permanent magnet, thereby inducing a flow of current in the ribbon. This device is called a pressure-gradient pickup because the motion of the ribbon is determined by the pressure difference at any point in time between its front and rear planes. This difference is proportional to the velocity of the air molecules that make up the sound wave, and, thus, this type of microphone is known by a third name, the velocity microphone.

Fig. 4-18. The ribbon microphone.

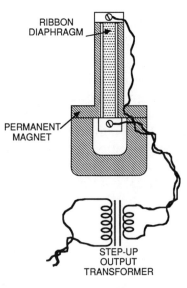

Since the ribbon is exposed to sound waves from both its front and rear sides, it is equally sensitive to sounds originating from both directions. Sound from the rear produces a voltage 180° out of phase with the voltage due to sound from the front. Sound waves that are 90° off axis will exert an equal, but opposing, pressure at the front and rear of the ribbon, with a net result of no displacement and therefore no signal output. As a result, the ribbon mic is inherently a bidirectional device, with a pickup pattern shaped like a figure eight. Other pickup patterns may be obtained when using a ribbon mic by redesigning the acoustic path to the rear of the ribbon to varying degrees. If the path to the rear of the mic is completely closed, the mic becomes an omnidirectional or nondirectional pressure-operated pickup. As the path is gradually opened, the pattern gradually becomes more and more heart-shaped (cardioid), then a narrow front lobe begins to develop (supercardioid) and finally, when the path is fully opened, the pattern is a figure eight.

Ribbon microphones are still in use today, with some vintage models regaining their original popularity, but only a relatively few modern designs utilize the ribbon element.

The Condenser Microphone

The *condenser microphone* (Fig. 4-19) operates according to an electrostatic principle rather than the electromagnetic principle governing the functioning of the dynamic and ribbon microphones. The head, or capsule, of a condenser microphone consists of two very thin plates (often on the order of 0.001 inch), one movable (diaphragm) and one fixed (backplate). These two plates form an electrical device known as a capacitor (also known as a condenser, and hence the name condenser microphone). A capacitor is capable of storing an electric charge. The amount of charge that is stored by a capacitor is determined by its ca-

Fig. 4-19. The condenser microphone.

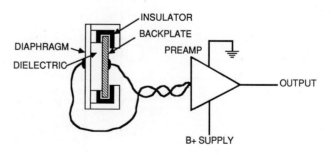

pacitance value and the voltage that is applied, according to this formula:

$$Q = CV \qquad \text{(eq. 4-1)}$$

where,

  Q is the charge, in coulombs,
  C is the capacitance, in farads,
  V is the voltage, in volts.

The capacitance of the capsule of a condenser microphone is determined by the composition and surface area of the plates (which are fixed values for a given set of plates), the dielectric, or substance between the plates (which is air and is also fixed), and the distance between the plates (which varies with the sound wave pressure). The plates are connected to opposite sides of a DC power supply, which provides a polarizing voltage for the capacitor. Electrons are drawn from the plate connected to the positive side of the power supply and forced through a high-value resistor onto the plate connected to the negative side of the supply. This continues to occur until the overall charge on the capsule (the difference between the number of electrons on the positive plate and the number on the negative plate) equals the capacitance of the capsule times the polarizing voltage. When this equilibrium has been reached, no further appreciable current can be made to flow through the resistor. When the diaphragm is displaced by a sound wave, the capacitance of the diaphragm changes polarity.

Referring to equation 4-1, it is obvious that Q, C, and V are interrelated. If the plate charge (Q) is set at a constant value and the diaphragm is displaced by a sound wave, the value for the capacitance (C) will be changed. In order for the equation to hold true, the voltage (V) must change in direct proportion to any change in the capacitance. The resistor and capacitor output circuit (Fig. 4-20) of the condenser

Fig. 4-20. Basic electrical schematic of a condenser microphone.

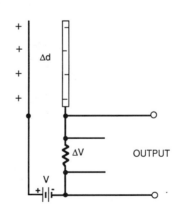

microphone, when placed in series with the power supply, will produce a voltage drop that is equal in magnitude to the supply voltage. When the voltage across the capacitor changes, the voltage across the resistor will change proportionately but in the opposite direction. This resistive drop is used as the output signal.

Since this output signal is of a low level and an extremely high impedance, a preamplifier must be placed within a short distance of the diaphragm (2 inches or less) and is often designed into the body of the mic to prevent hum, noise pickup, and signal level losses.

Because condenser microphones generally have built-in preamplifiers, many of them have a built-in attenuation pad (−10 dB or −20 dB) immediately following the capsule output and preceding the preamp input. This pad prevents overloading of the preamp if the diaphragm is exposed to very high pressure levels. If the capsule output level did get high enough to overload the preamp, no corrective action at the console would be able to fix the distortion. The use of an attenuation pad in a condenser mic avoids such a problem and should be considered necessary if the sound to be picked up is unusually loud and/ or is to be miked at a short distance (close-miked).

The Electret Condenser Microphone

The field of audio production for video has been greatly affected by a condenser-style transducer device, known as the *electret condenser microphone* (Fig. 4-21). The electret condenser mic basically operates according to the same principle as does the condenser mic. However, instead of requiring an external polarizing voltage to provide the required charge on the capacitor, the electret condenser element has the charge permanently stored within the capsule plates, in the form of a permanent electrostatic charge. This means that no external voltage supply is required to power this element. The high-impedance output from the diaphragm will, however, still necessitate the use of a follow-

Fig. 4-21. Basic electrical schematic of an electret condenser microphone.

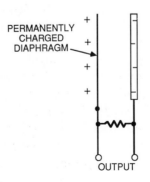

PERMANENTLY
CHARGED
DIAPHRAGM

OUTPUT

ing preamplifier. With today's integrated circuitry, this preamp can be readily designed to be powered internally from a 1.5-volt battery.

Polar Response and Sensitivity

The *polar response*, or *directivity*, refers to the sensitivity of a specific microphone type or model at varying angles (in a 360° circle) with reference to the front face of a microphone (0°). *Sensitivity* refers to the output of a microphone for a given reference acoustical input signal. It is possible for a mic to be designed so that the output for a given signal originating from the front (at 0°) is greater than that for an equal signal originating from the rear or side (at 90° or 180°).

The polar axis of a microphone (Fig. 4-22) may be viewed as an imaginary horizontal line drawn perpendicular to the front plane of the diaphragm. For a directional, or cardioid, microphone, signals entering the microphone from the front and within 55° of the polar axis on either side are said to be within the *on-axis acceptance angle* (of 110°), and signals originating from the remaining 250° may be said to be within the *off-axis rejection angle*.

Fig. 4-22. Polar axis and acceptance and rejection angles for a cardioid microphone.

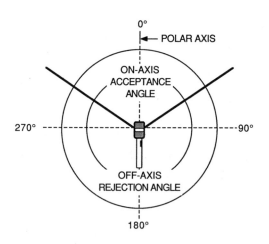

The polar response exhibited by a basic pressure-operated microphone is inherently omnidirectional (Fig. 4-23). That is, the microphone is equally sensitive (at a given frequency) to sound waves originating from any direction. In fact, the actual acoustic signals may only be picked up from the front of the microphone, so all incident sounds are combined at the diaphragm, regardless of their direction.

A pressure microphone may also be designed to have a directional response. The most common of these devices exhibit the *cardioid pat-*

Fig. 4-23. Om-
nidirectional
polar response
pattern.

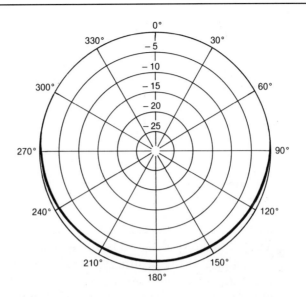

tern, whose name is derived from its heart shape (Fig. 4-24). In a car-
dioid-style directional microphone, the diaphragm is accessed at the
rear by an acoustical phase-shifting network, which is designed to delay all signals originating from the rear to compensate for the time it
takes for sound to travel from the rear port to the forward-facing diaphragm. In Fig. 4-25A, an on-axis signal (originating from the front of
the microphone) displaces the diaphragm. This signal also travels the
distance to the rear port and in doing so is displaced by 90° of a cycle.

Fig. 4-24.
Cardioid
(unidirectional)
polar response
pattern.

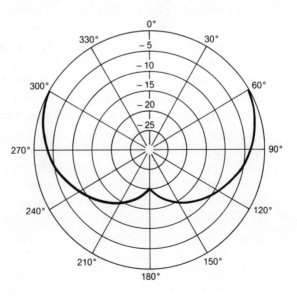

Fig. 4-25. Directional response of a pressure microphone.

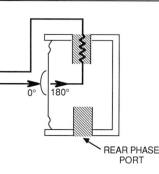

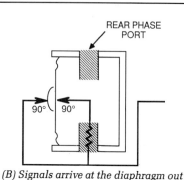

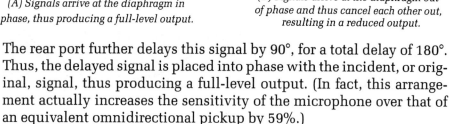

(A) Signals arrive at the diaphragm in phase, thus producing a full-level output.

(B) Signals arrive at the diaphragm out of phase and thus cancel each other out, resulting in a reduced output.

The rear port further delays this signal by 90°, for a total delay of 180°. Thus, the delayed signal is placed into phase with the incident, or original, signal, thus producing a full-level output. (In fact, this arrangement actually increases the sensitivity of the microphone over that of an equivalent omnidirectional pickup by 59%.)

Fig. 4-25B shows the same cardioid microphone system with the signal originating from the rear. In this case, the sound first enters the rear port, where it is delayed by 90°. Meanwhile, the signal also travels the outer acoustic path to the front of the microphone; covering this distance also results in a 90° phase shift of the signal. Thus, the delayed and nondelayed signals reach the diaphragm at 180° out of phase with each other, resulting in little or no response at the output of the microphone. Although total cancellation is not possible, attenuation figures between 20 dB and 30 dB can be expected.

As the directional microphone is rotated about its axis, the signal entering the rear port will partially cancel and partially reinforce the signal originating at the front of the diaphragm, thus producing the well-known cardioid pattern. The amount of rear attenuation displayed by a cardioid mic with respect to the on-axis signal level is known as the *front-to-back discrimination*.

Other response patterns within the cardioid family are known as the *supercardioid pattern* (Fig. 4-26A) and the *hypercardioid pattern* (Fig. 4-26B). Both of these patterns sacrifice good front-to-back discrimination in favor of a narrower pickup pattern. The maximum rejection angles for the supercardioid microphone are at 150° and 210°; the hypercardioid microphone is designed for maximum rejection at angles of about 120° and 240°.

The ribbon microphone gives the best example of the *figure-eight*, or *bidirectional*, *pattern* (Fig. 4-27). With the figure-eight pattern, a sound wave arriving from either 0° (on axis) or 180° (off axis) displaces

Fig. 4-26. Sub-types of the cardioid polar response pattern.

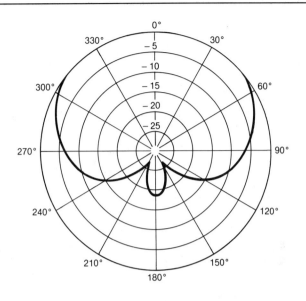

(A) The supercardioid pattern.

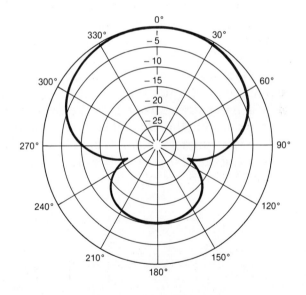

(B) The hypercardioid pattern.

the diaphragm fully, thus allowing for the maximum degree of output, or sensitivity. However, sounds arriving from the sides of the ribbon (from 90° or 270°) will set up equal and opposite pressure fronts at the diaphragm, which will cancel each other out in net force, resulting in a reduced output.

Fig. 4-27. The
bidirectional
(figure-eight)
polar response
pattern.

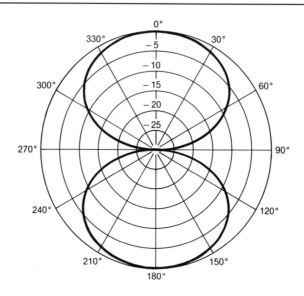

The condenser microphone is pressure-operated and is therefore, by nature, an omnidirectional device. Directional patterns may be achieved for a condenser mic, however, through the use of a perforated backplate, which operates in conjunction with a second diaphragm plate built into the capsule itself. This type of design is called a *dual-element condenser microphone* (Fig. 4-28). By placing two movable plates on opposing sides of the fixed plate, two capacitors are formed. By polarizing the two movable plates with a single voltage source and by polarizing the fixed plate with a source of the opposite polarity, an omnidirectional pickup pattern is achieved. If the polarization present at one of the diaphragm plates is decreased while holding the polarization of the other plate unchanged, the pickup pattern becomes in-

Fig. 4-28. Basic
electrical
schematic of a
dual-element
condenser
microphone.

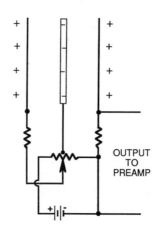

creasingly cardioid. When the movable plates are opposingly polarized and the fixed plate is set at ground potential, a figure-eight pattern results.

Many modern professional microphones allow selection of a pickup pattern by means of a switch located on the outside of the microphone's body. Certain new and older dual-element condenser designs, however, offer the ability to vary the polar response pattern continuously from omnidirectional to figure eight, either at the microphone body or from a remote control.

Proximity Effect

One property that is inherent in directional microphones is their exhibition of an increase in bass response as the sound source is brought closer to the capsule. This phenomenon, known as *proximity effect*, is most noticeable when the source is brought closer than 2 feet to the mic and increases proportionately as the distance decreases. Proximity effect is due to a pressure gradient that builds up at the front of the diaphragm and cannot be compensated for by the rear port at lower frequencies; the port is able to phase-shift correctly at higher frequencies but not at lower frequencies. At close distances, therefore, the overall frequency response breaks down, even when the source is directly on axis (Fig. 4-29). This boost of the bass response is found to be somewhat greater with figure-eight microphones than with cardioid ones. In order to compensate for this effect, a bass rolloff switch is often

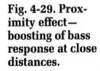

Fig. 4-29. Proximity effect— boosting of bass response at close distances.

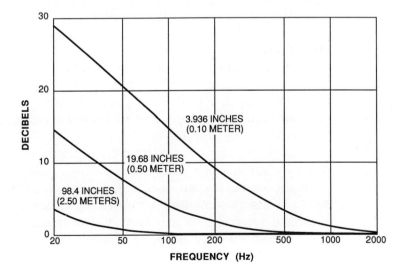

provided to bring the bass response back to the balanced position. In microphones designed for close working conditions, the frequency response may be internally rolled off at the low end and the proximity effect used to restore the correct low-end response.

Another method for reducing proximity effect, and the associated "popping" of the letters P and B, is to replace the directional microphone with an omnidirectional one for close-miked applications within the sound stage, studio, or on-location shoot.

Specifications

Specific data exists that may be of assistance in making the right choice of a microphone for a particular application. Some of this technical information may be supplied in the manufacturer's specification sheet; however, other vital information is often not included. The best gauge of a microphone's performance will, in the long run, be the overall sound produced through the device.

Frequency Response

The *frequency response* of a mic may give valuable information such as clues as to how the microphone will react at specific frequencies. Certain types of microphones may exaggerate high frequencies, and others, the middle or lower frequencies. A microphone that is able to respond at its output equally to all input frequencies is said to exhibit a *flat frequency-response curve*. Each type of microphone will be a good choice for some specific applications.

Transient Response

A significant piece of data concerning microphones, which presently has no accepted standard of measurement, is the *transient response*. Transient response is a measure of how quickly a diaphragm will react to a waveform (Fig. 4-30). This varies widely among microphones and is a major factor in sound quality.

The diaphragm of a dynamic microphone may be quite large (up to 2-1/2 inches in diameter). With the added weight of a coil of wire and core, the total mass can be very large, especially when compared with the total energy contained in the driving force of a sound wave. For this reason, a dynamic mic may be very slow in reacting to a waveform; its mass presents a large degree of resistance to motion. This physical fact

Fig. 4-30. Transient response of three types of microphones.

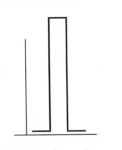

(A) Original waveform.

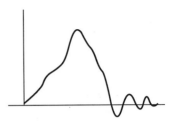

(B) Transient response of dynamic microphone.

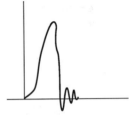

(C) Transient response of ribbon microphone.

(D) Transient response of condenser microphone.

tends to give an overall rugged, or gusty, character to the sound produced through a dynamic mic.

The diaphragm of a ribbon microphone, by comparison, is much lighter, being only a thin, corrugated ribbon. This type of diaphragm is able to react more quickly to a waveform, resulting in a cleaner and clearer sound. In their design, older models of ribbon mics often incorporate a large grill, whose function is to protect the diaphragm against sudden shocks or wind. These grills tend to set up an internal cancellation at higher frequencies, giving these mics a mellow sound quality for which they are often sought out.

The condenser microphone, on the other hand, is designed with an extremely light diaphragm, with a diameter between 0.25 and 2.5 inches and a thickness of about 0.0015 inch. Such a diaphragm is able to offer little mechanical resistance to the incident sound wave, and therefore a crisp and highly accurate sonic reproduction occurs over the entire audible range.

Directionality

A significant piece of data for the determination of a microphone's overall suitability for an application is its directional characteristic. Assertions concerning directionality may be meaningless without a plotted curve showing the frequency versus the off-axis response. All directional mics will exhibit some difference between the on-axis frequency-response curve and the off-axis one. However, well-designed models are able to maintain smooth off-axis characteristics over a wide acceptance angle. Directional information is generally given as a circular polar pattern, with coded curves that correspond to varying off-axis frequencies. The ideal directional microphone should have the same frequency response at all symmetrical on-axis and off-axis angles.

Sensitivity and Self-Noise Level

The sensitivity and self-noise level of a microphone serve to determine the overall signal-to-noise ratio. The higher the output level of a unit, the less amplification is required in a console's microphone preamplifier stage. By employing a variable-gain microphone preamp, the gain within the audio recording chain may be optimized for the best signal-to-noise ratio, so as to avoid the unnecessary amplification of thermal noise. The sensitivity of a microphone is determined through measurement of its output as delivered into a required impedance load at a specific sound pressure level. The most common sound pressure level used in the testing of a microphone system is 10 dynes/cm^2 (10 microbars or 74 dB) at 1 kHz, and the sensitivity rating is given in decibels relative to 1 milliwatt (stated as dBm).

The self-noise level of a microphone is the noise level generated by the unit when no signal is present (microphones are often tested for this under anechoic conditions). In a dynamic or ribbon mic, self-noise is generated by electrons moving within the coil or ribbon. In a condenser mic, self-noise is mainly due to the noise generated by the built-in preamplifier. Although recent strides in digital recording have forced improvements in microphone preamplifier designs, for any given device this noise cannot be totally eliminated, and often places restraints upon the overall dynamic range of the pickup.

Harmonic Distortion

Probably the least important specification for recording and pickup is the level at which a microphone will begin to create harmonic distortion. This is relatively unimportant because all modern, high-quality units are able to withstand intense sound pressures without creating appreciable harmonic distortion. This rating is usually given as dB spl (sound pressure level) at a total harmonic distortion (thd) of 0.5%.

Output Impedance

The *output impedance* of microphones varies from one design or type to another and may often be internally selectable. The output impedance rating is used to match the signal-providing capability of a microphone with the signal-drawing (input impedance) requirements of another device. This rating, which is given in ohms, may also be denoted by the letter Z. Most commonly used microphones have output impedances of 50 ohms, 150 ohms, 250 ohms (low Z), or 20K to 50K ohms (high Z). Each impedance range has its own particular advantages and disadvantages.

In the past, high-impedance mics (20 to 50 kilohms) were less expensive to use because of the high input impedance of tube-type amplifiers; that is, if a low-impedance mic was used, an expensive input transformer was required. (All dynamic microphones are, however, low-impedance devices by nature, requiring a step-up transformer to achieve a high output impedance.) A major disadvantage to the use of a high-impedance microphone lies in its high susceptibility to the pickup of surrounding electrostatic noise, such as that caused by fluorescent lights and electric motors. This makes the use of shielded ground cable (single-conductor cable) necessary. However, the use of a conductor surrounded by a shield creates a capacitor across the output of the mic. As the length of the cable increases, this capacitance also increases. At about 20–25 feet, the cable-created capacitance begins to short out much of the high-frequency information picked up by the microphone. Therefore, for best results with a high-impedance mic, it should generally be limited to use with a cable length of 25 feet or less.

Microphones with a very low impedance (50 ohms) have the advantage that their transmission lines are fairly insensitive to electrostatic noise. They are, however, somewhat likely to pick up induced hum due to electromagnetic fields, such as those generated by AC power lines. This unwanted pickup may be eliminated by the use of a twisted pair (a cable made up of two insulated conductors). Transmission lines for these 50-ohm mics do not need shielding, as this is effective only against electrostatic and not against electromagnetic pickup. Cable lengths are limited to about 100 feet, as a result of signal power losses due to excessive cable resistance. These losses, however, do not degrade the frequency response, but merely worsen the signal-to-noise ratio; they may be reduced through the use of larger conductors that exhibit less internal resistance.

The output impedances of the 150-ohm and 250-ohm mics have been found to suffer low signal losses, allowing for cable lengths of up to several thousand feet. As a result the 150-ohm or 250-ohm mic with a shielded twisted-pair has been found to offer the lowest attainable signal-to-noise ratio when used in conjunction with a transformer or transformerless balanced input circuit.

Balanced Transmission Lines

With a *balanced transmission line* (Fig. 4-31A), the signal current of a microphone is carried by way of two conductor wires. Neither of these leads are directly tied to the ground for the system, which is a third and electrically separate wire or outer shield. When an electrostatic or electromagnetic field impinges on the two audio leads, a current is generated and is in phase at both leads simultaneously. These equal charges will also arrive at the input transformer or balanced amplifier input in phase with each other, resulting in the nearly total cancellation of these unwanted interference signals at the input, while the alternating audio signal is allowed to pass unaffected.

Fig. 4-31. The microphone cable circuit.

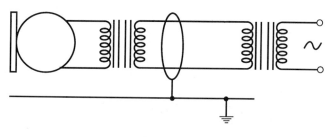

(A) The balanced line.

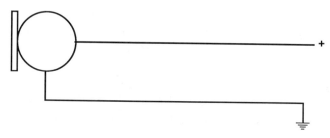

(B) The unbalanced line.

Phantom Powering System

Currently, most condenser microphones used in recording studios obtain their required DC polarizing voltage directly from the console by way of a *phantom powering system* (Fig. 4-32). Phantom powering is able to provide power to all microphones used as inputs to the console through the regular XLR connector and balanced cable, without any additional or external connections. In this system, the positive side of the voltage rail (B+) is fed to both microphone audio leads through a pair of resistors identically matched in value, while the negative side of the supply is connected to the shield and/or ground wire. A condenser microphone that is designed to make use of the phantom powering system no longer needs internal batteries, external battery packs, or individual AC-operated power supplies in order to operate.

Fig. 4-32. The phantom powering system is used to provide a voltage supply to the modern condenser microphone.

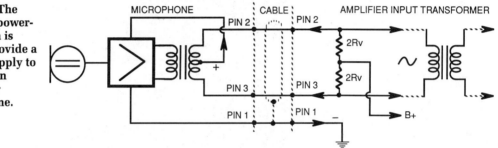

The phantom powering system does not interfere in any way with the normal operation of a dynamic microphone or an electret condenser microphone, since the positive voltage rail is cancelled at the microphone's input transformer. However, many professional consoles and mixers are equipped with an on/off switch controlling phantom powering at each input. This is because phantom powering presents a possible hazard to ribbon microphone elements in that it may cause destruction of a diaphragm. Therefore, when using a ribbon microphone, it is wise to switch the phantom powering system off at the specific input channel.

Direct Box

When a feed is required from an electrical musical instrument, such as an electric guitar or synthesizer, it is possible to record or pick up the instrument's sound directly, without the use of a microphone. The output of the instrument may be plugged directly into the recording console or mixer through a line input, or, if the instrument is being recorded on a sound stage or in the studio, a *direct injection (D.I.) box*, or *direct box* (Fig. 4-33) may be used. In the latter case, the output of the instrument is plugged directly into the input of the direct box, and the microphone output is plugged (via an XLR connector) into the console and then mixed or assigned to a track in the normal fashion for microphone input.

Fig. 4-33. Type 85 FET direct box. *(Courtesy Countryman Associates, Inc.)*

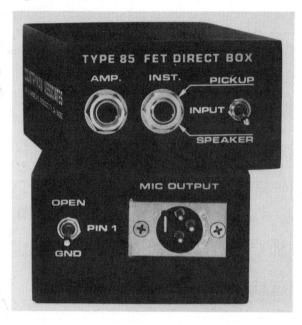

The direct box serves two purposes: it reduces the line level of an instrument to microphone level for ease of processing in a studio console during production; and, if necessary, it electrically isolates an instrument by means of an internal transformer or preamplifier, thus reducing the possibility of building up ground-loop potentials. In either case, the musician may monitor the instrument either by using headphones hooked up to the console's monitor circuits or by means of an instrument feed on the direct box, which may be used to feed the instrument's amplifier. The latter setup allows the option of placing a microphone at the amplifier and then mixing the direct and the miked signals together at the console or during the mixdown stage.

Two basic types of direct boxes are in general use today: transformer and active (or transformerless). The first type isolates and steps down the signal through the use of a transformer, a device that is often associated with a degraded transient response (or "frequency ringing"). The active direct box also serves to isolate the signal electrically, but without adding any appreciable sound of its own.

Since an amplifier lies at the heart of the direct box, it must be powered either by batteries or, in certain cases, by phantom powering.

Miking Techniques

The microphone and its associated techniques are the major tools of the sound engineer and recordist. There are many different types of microphones, and even a minor difference in basic design may give a mic a different and totally distinctive sonic character. Therefore, it is the responsibility of the audio professional to choose the right one for the application at hand. This, however, is really only half the story— choosing the right mic is critical for getting just the right sound, but microphone placement is equally as important.

It must be made clear at the start of this discussion that microphone placement is an engineer's tool, and what is considered bad technique now may be standard five years from now. As new styles of music and equipment develop, new sound recording techniques will also evolve, giving the recorded sound a new character. In light of these facts, it is wise to remember a recording adage that has been repeated for years and is known as the first rule of recording:

First Rule of Recording: There are no rules.

Sound recording is an art form and, as such, should be totally open to change and experimentation. It is this quality that keeps the music and audio industry alive and fresh. However, we can add to the above fundamental rule, in reference to microphone techniques:

There are no rules, only guidelines.

This section presents a few of these guidelines in light of recent advances and adaptations to the field of audio production for video.

Distant and Close Miking

In modern studio and sound stage recording, there are basically two general techniques for microphone placement: distant miking and close miking.

Distant miking refers to the positioning of one or more microphones at a distance of five or more feet from the sound source that is to be picked up. This technique may serve two functions: (1) to place the mic at such a distance that the entire sound of a musical instrument or ensemble is picked up, thus preserving the overall tonal balance of that instrument or ensemble (Fig. 4-34A) and (2) to place the mic such that the acoustic environment is picked up also and is thus combined with the sound source's direct signal (Fig. 4-34B). Distant miking is often used with large instrumental ensembles (such as a symphony orchestra) or choral ensembles and thus relies heavily on the integrity of the acoustic environment (such as a concert hall). In such a situation, the microphone(s) is placed at such a distance as to strike a balanced pickup between the ensemble and the environmental acoustics. The use of distant miking is not limited to the pickup of classical music, as is sometimes believed. It may be used in many different circumstances, ranging from the pickup of overall crowd voices to the pickup of the larger-than-life, driving sounds of a drum in a large multitrack recording studio. Used for its effect, distant miking can add depth to a sound. For example, if a sound is played loudly over the studio or sound stage monitors and then recorded through a distant pair of mics, the result may be a fatter, deeper sound, either alone or when remixed with the original source.

Fig. 4-34. Examples of distant miking; (A) Miking at a distance of 10 feet, and (B) Miking at a distance of 30 feet.

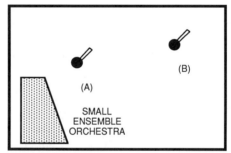

Close miking refers to the placement of a microphone in close proximity to a sound source, at 1 inch to 3 feet (Fig. 4-35). This technique is most often used in the modern multitrack recording studio, as well as in audio-for-video production for ENG and EFP. The miking of a sound source at such a close range will effectively exclude the acoustic environment from being recorded onto tape. The intensity of a sound diminishes with the square of the distance as one moves away from the source. Therefore, a sound originating at a distance of 6 feet from a mic will be insignificant compared to the same sound at 3 inches away.

With close miking, only the desired on-axis sound is recorded onto tape. The acoustical environment will, for all practical purposes, not be picked up.

Fig. 4-35. Example of close miking.

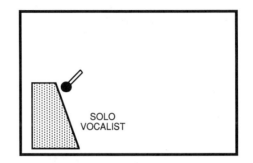

Close miking techniques are very important in audio production for video. In the recording studio, external noises are kept to a minimum, but the sound stage or EFP video shoot is often plagued by extraneous noises (originating from equipment, personnel, street traffic, etc.). Close miking can exclude most of this unwanted sound.

One problem associated with the use of close miking techniques in video production is that of keeping the microphone out of the camera range. In situations where the mic must not appear within the picture, a boom- or fishpole-mounted, highly directional microphone may be employed. As an alternative to this, a clip-type microphone may be worn by the on-camera talent, or a pickup device may be hidden on the set as close to the talent as possible.

In the modern recording studio or sound stage, the basic idea underlying multitrack recording is to maintain control over individual tracks, making the isolation of recorded tracks highly important. If extraneous sounds (instrument or voice) are picked up by a mic that is recording a nearby source, a condition known as *leakage* has occurred (Fig. 4-36). Since the mic would be picking up both a direct and indirect signal (leakage), control over the desired signal on the recorded track in the mixdown phase would be difficult to achieve without also affecting the level and sound character of the indirect source. Excessive leakage will tend to make a soundtrack more like live sound and less intelligible. Needless to say, unwanted leakage is to be avoided; however, even when its effect is desired, caution is necessary since the recorded tracks can easily take on the negative aspects of leakage (leading to loss of control in the mixdown phase).

There are two ways of correcting a leakage problem such as that illustrated in Fig. 4-36. Placing the two microphones closer to their respective sources and/or placing an acoustic barrier known as a flat between the sources will serve to reduce the effects of acoustic leakage.

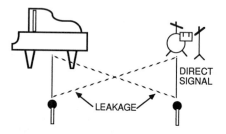

Fig. 4-36. Leakage may occur from one microphone to another improperly placed pickup.

With close miking, the total sound balance of an instrument may not be picked up. The mic may be placed so close (at 1 to 6 inches) that only a small portion of the instrument's sound may be picked up, producing a limited, or area-specific, balance. At such close distances, the movement of the mic by only a few inches may change the tonal balance for the instrument.

As the first rule of recording implies, the choice of close or distant miking is up to the sound engineer. A jazz flute can be made to sound interesting whether it is recorded in the studio at a distance of 1 foot with added reverb or in the Taj Mahal at 50 feet. Through experimentation, in many cases, the mix can yield very interesting effects.

Phasing

Stereo effects may be produced through the use of two microphones to pick up the sound from a single source, but care must be taken to make certain that these two pickups are electrically and acoustically in phase. If this is not the case, signal cancellations may occur at various frequencies, changing the volume and overall character of the sound. In the case where out-of-phase microphones have been placed left and right in a mix, a listener to stereo will hear the desired balance, while a listener to mono (over a car radio, for example) will hear very little of the recorded instrument. One simple check for phasing is to assign the output from both (or all) of the mics to one speaker and listen for drastic changes in the sound. If one or more mics are 180° out of phase with respect to the others, the problem may be that a cable was miswired (assuming that the studio equipment in use was properly checked for phase integrity when patched or wired up). This may be solved by either replacing the cable with a good one or repairing the cable by correcting the polarity of the audio leads (pin 2 and pin 3 on the XLR connector), by using a phase-reversal adapter, or by reversing the phase on the console's input strip (where possible).

Acoustical cancellation problems may also be produced when too many microphones are used at too close a distance to a source, allowing for excessive leakage (Fig. 4-37). Differing path lengths from the source to the microphones will often result in differences in phase between the microphone outputs for a given sound. If the two signals are combined at any point, such as in the mixing stage, phase cancellations will occur, producing severe dips in the overall frequency response. In order to avoid these phase cancellations, the *three-to-one principle* should be employed whenever microphone outputs are to be mixed together. This principle states that the distance between any two microphones should be at least three times the distance from either of them to the source that they are picking up (Fig. 4-38).

Fig. 4-37. Phase cancellation problems may result when close miking is used with a multiple-microphone setup, such as for a modern rock drum set.

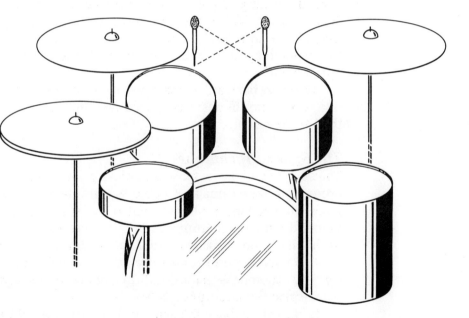

Stereo Miking Techniques

As stereophonic audio reproduction techniques become the norm in audio production for video, the need for an understanding of stereo miking techniques will also increase. Here stereo miking techniques

Fig. 4-38. The
three-to-one
principle for
miking.

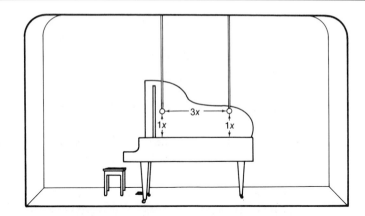

refers to the use of two microphones for obtaining a stereo sound image, as opposed to single-point miking, with which a stereo image is created at the audio production console through panning techniques.

Stereo miking techniques may be used for either close or distant (overall) miking of a large or small ensemble or a single instrument recorded in an on-location or studio environment. These techniques may also be applied to ENG or EFP production. The only limitations for the engineer or sound recordist are those placed by imagination in application.

There are basically three types of stereo miking techniques that utilize two microphones: the spaced technique, the X/Y technique, and the M/S technique.

The Spaced Technique

The *spaced technique* for sound pickup was among the first known to produce a stereo image. Generally, this technique employs two (or more) matched microphones, which are set symmetrically along a line perpendicular to the front plane of the sound source. The polar pattern of this pair, their spacing, and their distance from the sound source may all vary. Stereo sound information is created in these configurations by differences in the amplitude of the sound waves and their time of arrival at the microphones.

In regard to phase integrity with the spaced miking technique, the three-to-one principle again applies. That is, the distance between the spaced pair should be at least three times the distance between each microphone and the source.

The X/Y Technique

The *X/Y technique* is a coincident technique, which consists of using a pair of matched directional microphones whose grills are placed as close together as possible without touching and which are set at an angle within a horizontal plane (Fig. 4-39). The angle formed by this microphone pair is typically symmetrical about the center line from the sound source and is between 60° and 120°. The specific angle chosen determines the "apparent width" of the stereo image.

Fig. 4-39. The X/Y technique for stereo miking.

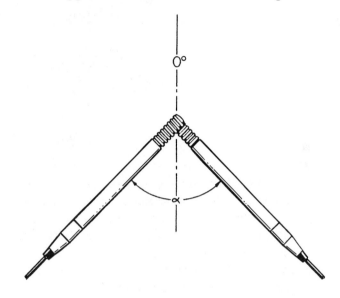

(A) Placement of microphone pair (α = 60°–120°).

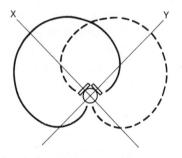

(B) Resultant polar response pattern.

There are microphones currently available from various manufacturers that embody the X/Y technique within their design. These stereo mics contain two coincident diaphragms mounted within a single housing (Fig. 4-40). The upper element may be rotated as much as 180°, accommodating all possible offsets. Since this type of system relies on the polar response patterns of a matched pair of pickups, considerable off-axis coloration is possible. Therefore, as for any stereo miking technique, these microphones should have as good a polar response as possible.

Fig. 4-40. The AKG C-422 stereophonic condenser microphone. *(Courtesy of AKG Acoustics, Inc.)*

The *Blumlein crossed figure-eights technique* (Fig. 4-41), named after the British scientist Alan Dower Blumlein, is the earliest of the X/Y techniques. Two microphones with figure-eight response patterns are oriented at 90° to each other. With this configuration, placement and surrounding acoustics are critical for maintaining a proper ratio of direct to reverberant sound. It is often commented that this technique produces a very natural sound.

The *binaural technique* is intended specifically for use when playback is to occur via headphones. This technique is often configured with two omnidirectional microphones placed in the ears of a dummy head (Fig. 4-42) or spaced 5–6 inches apart with a sound-absorptive baffle between them to simulate the sound received at the ears of a listener. This technique often proves to produce sound that is quite realistic, with good illusions in both horizontal and vertical planes when

Fig. 4-41. The Blumlein crossed figure-eights response pattern.

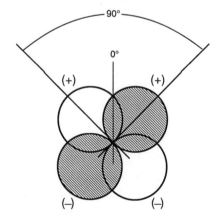

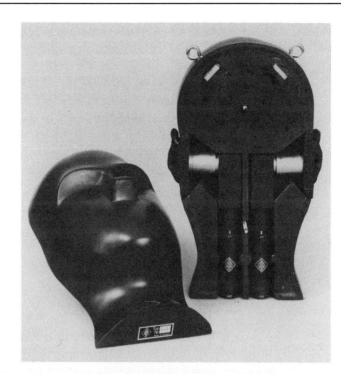

open-air headphones are used. Unfortunately, these characteristic illusions do not reproduce as well over loudspeakers.

The M/S Technique

The *M/S, or mid side, technique* is similar to the X/Y technique in that two diaphragms positioned within close proximity may be used, as in a stereo microphone system, or two coincident microphones of the appropriate polar patterns may be employed. In a classic M/S system (Fig. 4-43), the microphone capsule designated as the mid (M) one will have a cardioid pickup pattern oriented in the direction of the sound source.

Fig. 4-43. The response pattern for the M/S technique.

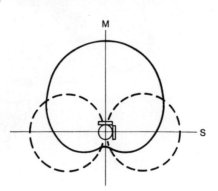

The other capsule, designated as the side (S) capsule, has a figure-eight response pattern, which is oriented laterally. Direct signal information is thus picked up by the mid capsule, while ambient and reverberant information is received by the side capsule. The outputs of the capsules are then processed by a sum-and-difference matrix, which resolves them into conventional stereo signals: (M + S) and (M − S).

One major production advantage of this type of system is its provision for absolute monaural predictability. Whenever the left and right signals are combined, the sum is solely the output from the mid component [(M + S) + (M − S) = 2M], which contains the direct signal information. Since it is generally desirable that there be less reverberation within a monaural signal than within a stereo one, there is a built-in advantage to the use of the M/S technique for monaural recording.

In stereo production, the M/S technique offers the sound engineer or recordist a great advantage in increased sound control. It is possible to vary the ratio of mid to side information delivered to the sum-and-difference matrix at the audio production mixer or console, allowing the direct-versus-ambient quality and the stereo width to be varied. It is equally possible, given two properly phase-aligned audio recording tracks, to record the mid information on one audio track and the side

Fig. 4-44. Audio Engineering Associates MS 38 active M/S matrix. *(Courtesy of Audio Engineering Associates.)*

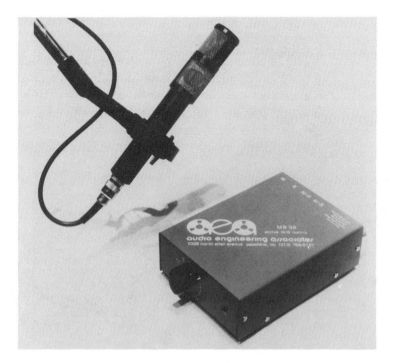

information on another, allowing the data to be mixed into an X/Y compatible signal at a later time, in the postproduction mixdown phase. An added advantage of this process is that important decisions regarding stereo width and depth may be made at a later, more controlled date.

One commercially available sum-and-difference matrix, which utilizes an active combining circuit, is Audio Engineering Associates MS 38 (Fig. 4-44). This line-level device allows for the adjustment of the mid-to-side ratio via a single knob. The MS 38 is intended to follow the microphone preamplifiers from the mid and side mics in a real-time recording situation or upon playback of these signals from tape during postproduction (Fig. 4-45).

The sum-and-difference matrix may also be configured as a dual-transformer arrangement (Fig. 4-46), which may easily be incorporated into a system through the patch bay or as a separate housing. The diagram in Fig. 4-46 shows the transformer as two Jensen JE-MB-D; however, any such device may be used that suits the given needs and quality standards (i.e., UTC-A-43, Triad A-67-J, etc.).

Fig. 4-45. The MS 38 function.

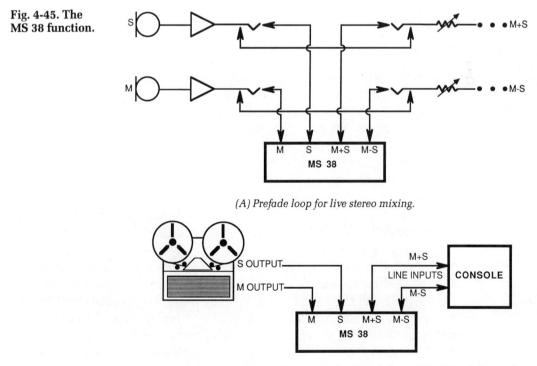

(A) Prefade loop for live stereo mixing.

(B) Postproduction mode for changes in the stereo image within the mixdown phase.

**Fig. 4-46.
Schematic for a
transformer-type
sum-and-
difference
matrix.**

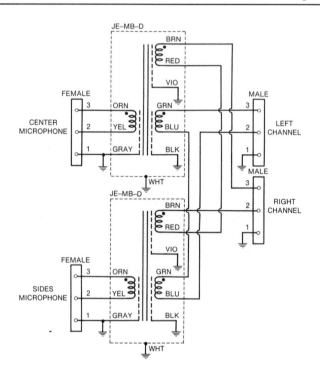

**Fig. 4-46.
Schematic for a
transformer-type
sum-and-
difference
matrix.**

The Accent Microphone

Commonly, an *accent microphone* is added to a basic two-mic setup within the overall stereo sound stage to give special emphasis to a vocal or instrumental soloist, through the placement of a monaural pickup element (in most cases) closer to the intended sound source.

When an accent microphone is used, care must be exercised as to both the placement and the type employed (as well as the amount of signal introduced into the mix), so that the mic contributes only a proper representation of the soloist and does not color or change the balance of the surrounding elements within the stereo perspective. The output of an accent microphone can often seem "out of context" with respect to the overall stereo sound since it is introducing into the mix sound from a closer miking distance, which precedes the sound from the basic microphone pair in time. One way of compensating for this time discrepancy is to delay the signal from the accent microphone (through the use of a digital delay unit), so it "arrives" in the mix at the same time as or slightly later (10−16 milliseconds) than the signal from the basic pickup. This signal should be introduced into the mix (for natural sounding reproduction) in order to add presence to the solo sound but not so it is perceived as a separate pickup in and of itself. Similarly, the proper panning of the accent sound into the overall stereo image is critical, or the result may be a wandering image with changes in solo intensity.

Special-Purpose Microphones

The Soundfield Microphone System

The *Soundfield microphone system* was developed by Calrec Audio Ltd., following theoretical work by Michael Gerzon, under the direction of the National Research Development Council in England. This multiple-pickup system employs four transducer elements, which are fitted into a single housing (Fig. 4-47A). Via electronic manipulation within the Ambisonic controller 47B (Fig. 4-47B) the picked-up signals are converted into "surround sound" signals, in addition to conventional left and right stereo signals.

In their original form, the four signals are derived from the four transducers, which are arrayed in a near-coincident tetrahedron. The transducers' outputs (the A-format signal) are electronically matrixed to produce (Fig. 4-48):

1. An omnidirectional component relaying the pressure of the sound wave at the microphone.
2. A pressure-gradient component relaying the vertical (up and down) information for the sound wave.
3. A pressure-gradient component relaying the lateral (left and right) information.
4. A pressure-gradient component relaying the lateral (fore and aft) information.

These virtually coincident signals are encoded into a B-format signal, which contains the overall lateral, vertical, and pressure information. The B-format signals may be stored on 4-track tape for postproduction use, or they may be processed immediately for resolution into quadraphonic, Ambisonic, or conventional stereo signals.

The electronic controls of the Soundfield microphone system also give the mixing engineer the ability to steer, pan, tilt, vary the included angle, alter the directional pattern, and otherwise change the overall stereo perspective and imaging. All this may be accomplished without altering microphone placement or even directly from tape during postproduction.

Fig. 4-47. The Soundfield microphone system. *(Courtesy of Audio + Design/ Calrec, Inc.)*

(A) Calrec Soundfield microphone.

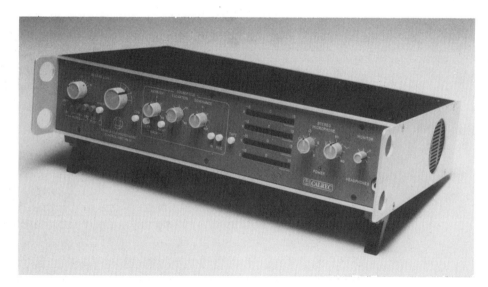

(B) Calrec Ambisonic MK IV control unit.

The Clip Microphone

The *clip microphone* is a small, high-quality, multiple-purpose micro-phone, often incorporating an electret condenser element. It is most often used where space is at a minimum and/or visibility is to be mini-mized. The clip microphone is able to fit into a wide range of applica-tions, but is most commonly found as an on-camera pickup for announcers, where it has all but replaced the standard lavalier micro-phone. Some examples of clip microphones are shown in Figs. 4-49, 4-50, and 4-51.

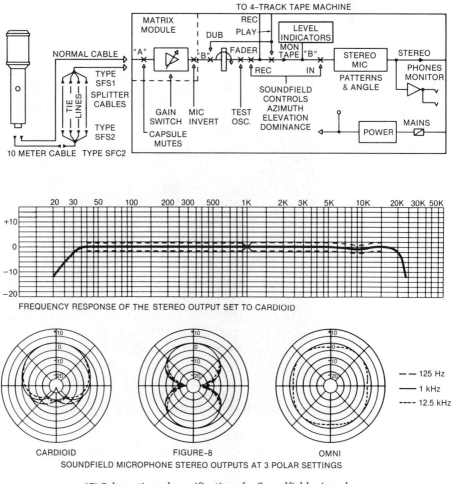

FREQUENCY RESPONSE OF THE STEREO OUTPUT SET TO CARDIOID

SOUNDFIELD MICROPHONE STEREO OUTPUTS AT 3 POLAR SETTINGS

(C) Schematic and specifications for Soundfield microphone.

Fig. 4-48. The XYZ axis of the pickup of the Calrec Soundfield microphone.
(Courtesy of Audio + Design/ Calrec, Inc.)

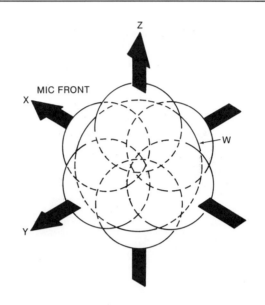

Fig. 4-49. Sony ECM-44 lavalier microphone.
(Courtesy of Sony Corporation of America, Inc.)

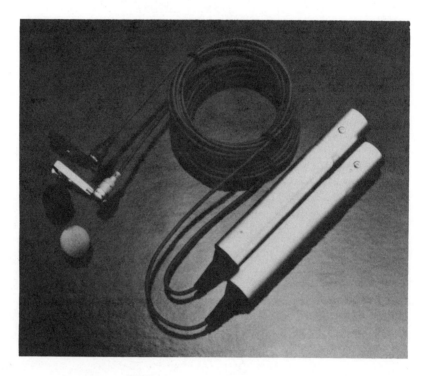

The frequency response of these mics is wide and smooth; however, the high frequencies are often emphasized in order to compensate for the bass boost that naturally occurs since clip mics are usually used for voice pickup and are therefore located in close proximity to the human chest cavity. On certain units, it is possible to flatten the response by making minor internal modifications.

All clip microphones have two sections: the pickup and the interface housing. The pickup is composed of the diaphragm, the pre-amplifier (often a single-stage FET transistor), and the unit housing. The interface housing is composed of the final preamplification stage (optional), voltage powering interface (either phantom or battery), output transformer, battery housing, and battery on/off switch (optional).

The standard options available with clip mics may include a universal mount, a tie bar mount for direct attachment to clothes, a tie tack mount, and a windscreen for outdoor use. Optional mounting devices are available for use with the Countryman flat-top and Crown clip microphones, including a drum mount, a cymbal mount, and a horn mount.

The Boundary Microphone

In recent years, another microphone that utilizes an electret condenser element has gained a great deal of popularity and recognition.Known by the generic name *boundary microphone*, this device is also known by the trade name given by the manufacturer (Crown International, Inc.)—Pressure Zone Microphone® or PZM® (Fig. 4-52). The theory behind the PZM microphone is that sound approaching a boundary (floor, wall, table, podium, etc.) creates a thin zone in which the direct and reflected signals are coherently in phase and reinforce each other. If a transducer is placed within this zone, the direct and reflected signals are treated as one signal with a flat frequency-response curve, which is unaffected by signal reinforcement or cancellation (Fig. 4-53). The polar response pattern of the boundary microphone is hemispheric in nature, and sound sources that move on the surface of this sphere will not alter the quality of the pickup. The low-frequency performance of the boundary microphone may, in most cases, be enhanced by placing it on a large flat surface (a table top or a piece of plexiglass up to 2 meters square works very well).

The boundary microphone has had considerable success in the recording studio due to its unique and various styles of pickup. It is also utilized in video and film production because of its inobtrusive size.

The Superdirectional, or Shotgun, Microphone

The *superdirectional microphone* (Fig. 4-54), also known as the *shotgun microphone*, is commonly used in double-system sound production for film and video. This is due to the extremely narrow acceptance angle of this type of microphone, which is able to accept signals that come in directly on axis while rejecting signals that are as little as 15° off axis at higher frequencies. A desirable effect of the narrow acceptance angle is the increased sensitivity, allowing the shotgun mic to pick up desired signals optimally while rejecting unwanted background noise.

Fig. 4-50. Countryman Isomax II. *(Courtesy of Countryman Associates, Inc.)*

(A) Isomax II clip microphone.

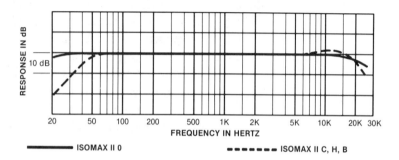

(B) Frequency response of Isomax II.

The Multiple-Capsule Condenser Microphone System

In recent years, the audio industry has seen the emergence of the *multi-capsule condenser microphone system*, which has many applications in the fields of music recording, film, video, and theater. This system combines a small, modular capsule with a microphone preamplifier housing. This type of system has two basic advantages:

1. The capsule may be replaced with another that has the polar and frequency response best suited to a given application.
2. The capsule may be made remote with respect to the preamplifier housing by means of an angled extension tube or

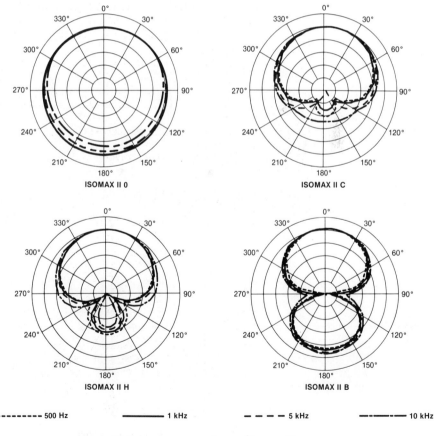

(C) Polar response patterns of Isomax II.

**Fig. 4-51. Crown
GLM-100/E.**
*(Courtesy of Crown
International.)*

(A) GLM-100/E clip microphone.

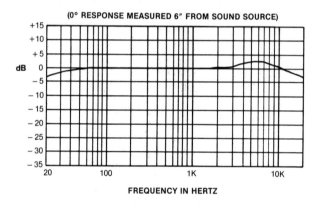

(B) Frequency response of GLM-100/E.

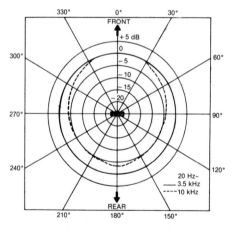

(C) Polar response pattern of GLM-100/E.

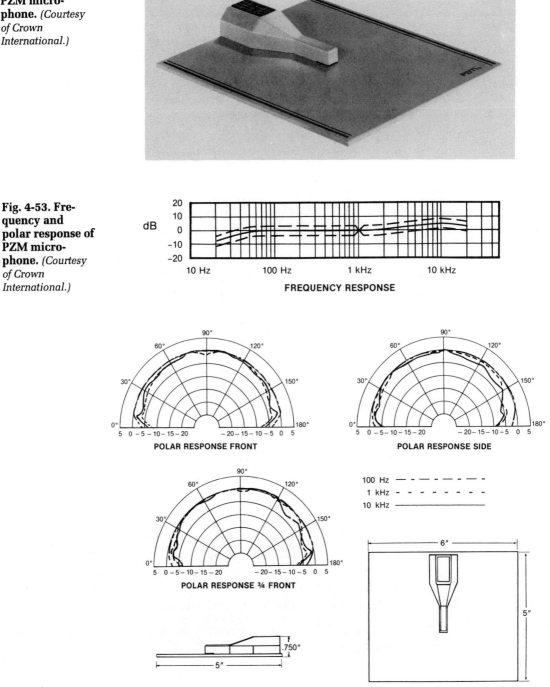

Fig. 4-52. Crown PZM microphone. *(Courtesy of Crown International.)*

Fig. 4-53. Frequency and polar response of PZM microphone. *(Courtesy of Crown International.)*

**Fig. 4-54. JVC
MU-6200E su-
perdirectional
microphone.**
*(Courtesy of JVC
Company of
America.)*

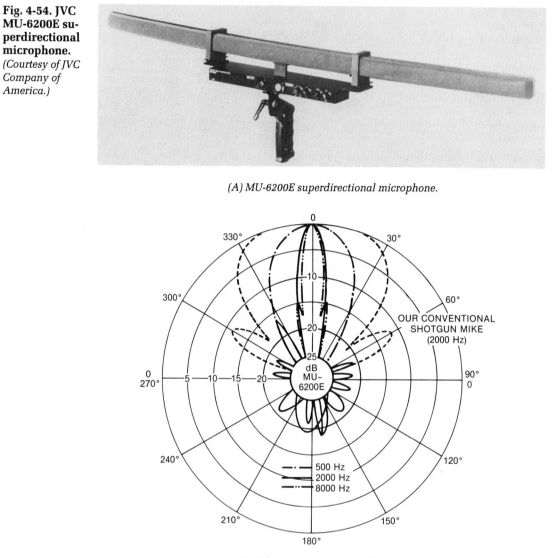

(A) MU-6200E superdirectional microphone.

(B) Polar response pattern of MU-6200E.

a flexible cable connection. This allows the capsule to be mounted or hung off-camera or in difficult to reach locations.

Figs. 4-55 and 4-56 show two examples of multiple-capsule condenser microphone systems.

Fig. 4-55. The AKG C-460, a multiple-capsule condenser microphone system. *(Courtesy of AKG Acoustics.)*

(A) C-460 and CK-61 ULS.

(B) CK-1 capsule and H-48 hang/stand adapter.

(C) CK-8X short shotgun capsule.

**Fig. 4-55 (cont).
The AKG C-460,
a multiple-cap-
sule condenser
microphone sys-
tem.** *(Courtesy of
AKG Acoustics.)*

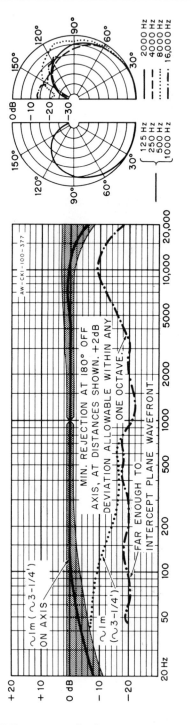

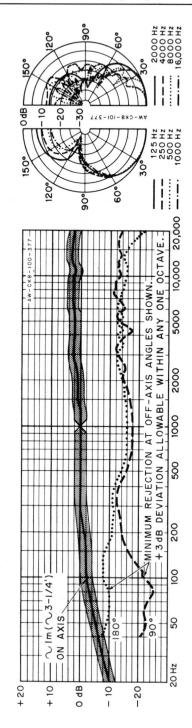

(D) *Frequency and polar response of CK-1.*   (E) *Frequency and polar response of CK-8X.*

The Boom and Fishpole

Many differing mechanical arrangements may be used in supporting a microphone out of the camera's line of sight. One such device is the *steerable microphone boom* (Fig. 4-57), which is often used in productions set in a sound stage when the movement of the on-camera talent prohibits the use of positioned or hidden microphones on the set.

These booms are generally made up of a drivable wheel base, allowing for mobility, and an extendable and retractable boom, which is also adjustable with regard to microphone direction. The microphone is attached to the boom by way of an elastic shock mount, which serves to isolate the pickup from vibrations originating in the floor or from movement. An extra precaution that may be taken so as not to defeat this isolation is not to use a stiff microphone cable, which may transmit the vibrations directly to the pickup capsule. Instead, a short, flexible cable should be employed at the mount in order to absorb these vibrations.

Fig. 4-56. The Calrec 2000 series multiple-capsule condenser microphone system. *(Courtesy of Audio + Design/ Calrec, Inc.)*

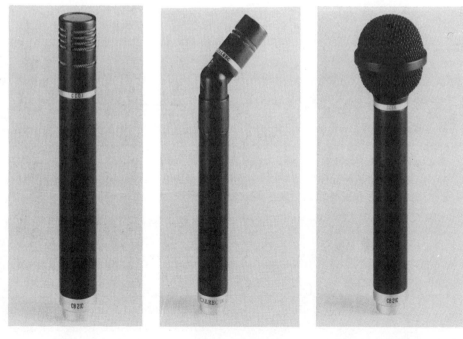

(A) CB 21C with CC01 capsule.   *(B) CB 21C with CK-10 capsule swivel adapter.*   *(C) CB 21C with CC56 capsule, a hand-held voice microphone.*

**Fig. 4-56 (cont).
The Calrec
2000 series
multiple-capsule
condenser
microphone
system.** *(Courtesy
of Audio + Design/
Calrec, Inc.)*

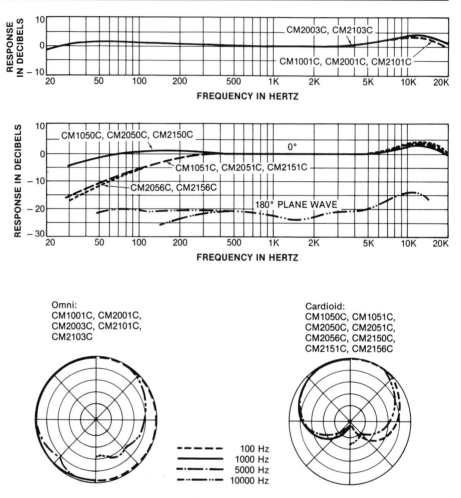

(D) Frequency and polar response of 2000 series.

Another form of boom is the *fishpole* (Fig. 4-58), which is often used in productions where movement is restricted to a rather small area (as in EFP). The fishpole is a hand-held, often retractable, pole with the microphone mount located at the end piece. It may be employed overhead or below the camera's line of sight for an on-camera pickup. In the latter case, care should be taken to avoid phase interference with the floor boundary in a sound stage or on-location setting.

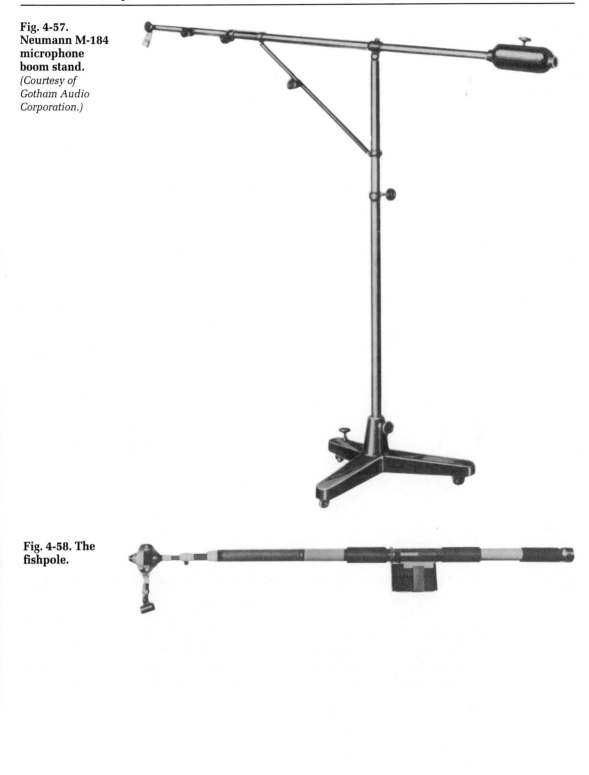

Fig. 4-57. Neumann M-184 microphone boom stand.
(Courtesy of Gotham Audio Corporation.)

Fig. 4-58. The fishpole.

5 *Audio Postproduction for Video*

The term *audio for video* has recently evolved to describe two components of audio production used in the video medium:

- High-quality stereophonic audio
- Multitrack production techniques

The adaptation of these components for the medium of video has given production teams the ability to further improve television through the introduction of creative, high-quality audio.

The increased demands for quality audio production for video have created a need for more technically and artistically complex processes. In order to meet this challenge, the recent shift within the video industry has been toward a greater emphasis on the stage of postproduction. *Postproduction* (also commonly referred to as "post") refers to the technical and creative manipulation of original and/or electronically generated source material in order to create a complete and continuous film, video, or audio program.

The Multitrack ATR in Audio for Video Production

In a television production studio, the concerns of the broadcast audio engineer are virtually the same as those of an "on-the-air" sound mixer. At present, the broadcast medium relies heavily on the continuation of the "live" format (in the form of talk shows, news and information programming, etc.). However, in a video production house, entertainment programming makes the heaviest demand on video sound production, and the recent upsurge in the popularity of high-fidelity, stereophonic sound has made the "live" approach increasingly ineffective. Given this fact, the introduction of multitrack production techniques into the audio-for-video process became highly advantageous.

The use of multitrack recording techniques during production phases on the sound stage or in the EFP video shoot allows the sound recordist to separate off-camera and on-camera dialogue, live sound effects, and sound presence onto individual tracks of a synchronized multitrack ATR. The advantages to using these techniques are twofold. First, multitrack production eliminates the need for the mixdown (sub-mix) of important sounds onto a few audio tracks in the field. The major drawback of such field mixing is that it often produces an improperly recorded or mixed balance. If a multitrack ATR is utilized instead, balances may be adjusted later, within postproduction, and improperly recorded tracks may have a better chance of being saved through equalization due to their multitrack isolation. The second advantage of multitrack audio-for-video production is the improvement in stereophonic sound that it brings. The field mixing of source audio to one or two tracks of an ATR or VTR results in the production of a basically monaural sound track. In multitrack field production, individual audio sources may be isolated and mixed at a later and more conducive time, producing a more realistic and involving final sound.

The Multitrack ATR in Audio-for-Video Postproduction

The role of the multitrack ATR is to provide increased flexibility in the production of quality stereophonic sound. Within the phases of audio-for-video postproduction, the ability to record and isolate multiple sound sources synchronously in time is a very powerful tool. Through the development and utilization of modern recording and synchronization techniques, it has become possible to manipulate sound sources independently in the postproduction phase, eliminating the stress and concerns associated with recording live sound.

In addition to the greater control gained from multitrack isolation of the sound elements, multitrack production yields the added advantage of allowing attention to be paid to individual sound sources throughout the stages of building a sound track. Three tools are made available to the sound engineer or sound mixer who utilizes multitrack production techniques: overdubbing, punch-in and punch-out, and track bouncing.

Overdubbing

One of the key work-saving features of the multitrack recording process is the capability for *overdubbing* (Fig. 5-1). In overdubbing, an original sound source may be recorded onto one or more tracks by the ATR, after which the tape may be recued to the desired starting position. Then, during sync mode playback of the originally recorded material (via headphones or speaker monitors), a second sound source may be recorded onto another track or tracks in a continuous fashion. This process is repeated until the complete soundtrack has been built up. If an error occurs in the building of tracks (in timing, level, etc.), it is generally a simple matter to recue the tape back to the desired in point and reexecute the overdub.

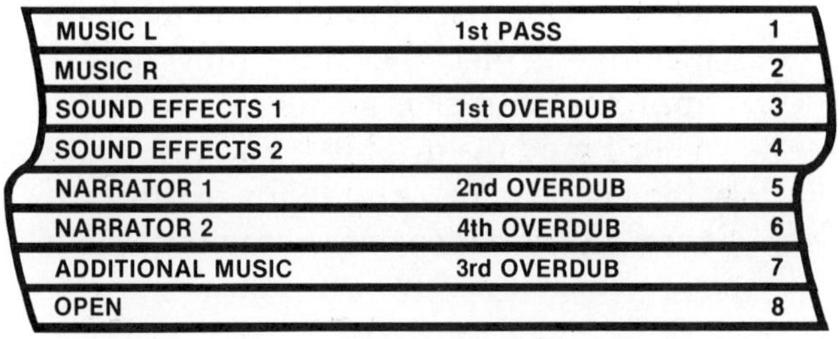

Fig. 5-1. With the multitrack overdubbing facility, music beds, jingles, and spots can be assembled track by track, eliminating the need for critical "real-time" cues. Timing can be worked out in the first passes and then levels adjusted later, in the final mix. *(Courtesy of Sony Corporation of America, Inc.)*

MUSIC L	1st PASS	1
MUSIC R		2
SOUND EFFECTS 1	1st OVERDUB	3
SOUND EFFECTS 2		4
NARRATOR 1	2nd OVERDUB	5
NARRATOR 2	4th OVERDUB	6
ADDITIONAL MUSIC	3rd OVERDUB	7
OPEN		8

Professional one-inch tape, eight-track format

Punch-In and Punch-Out

In multitrack production of music and certain sound beds, *punch-in* and *punch-out* (Fig. 5-2) have become established as important features. If a recorded track contains an error or musical flaw, it is possible

**Fig. 5-2. Punch-
in enables the
operator to re-
place material
selectively and
correct mistakes
prior to the final
mixing process.**
*(Courtesy of Sony
Corporation of
America, Inc.)*

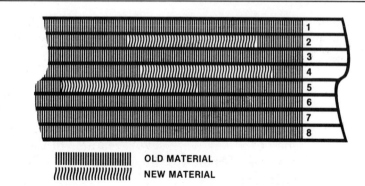

Fig. 5-2. Punch-in enables the operator to replace material selectively and correct mistakes prior to the final mixing process. *(Courtesy of Sony Corporation of America, Inc.)*

to rerecord acceptable sound for this specific section, thereby correcting the problem without the need for a complete retake.

In the process of punching, all tracks are monitored off the record head of the ATR while it is in the sync mode. The track (or tracks) on which something is to be changed is switched into the record ready status, while the input signal is monitored (many professional multitrack ATRs are able to perform input/reproduce switching automatically when placed in the sync mode). When it is time to rerecord, the machine is placed into the play mode. Then, at the desired point, the appropriate track is switched into record. At the end of the section, the record mode is disabled, and the ATR track will return to reproduce. Upon successful completion of this process, the recorded program will show no evidence of the previously existing flaw.

Track Bouncing

Track bouncing is a production technique often used in multitrack recording to expand the number of available tracks in order to allow for additional overdubbing. In Fig. 5-3, the first six tracks of a 1-inch, 8-track tape were "filled" with recorded material, leaving two available, or open, tracks. On a second pass of the tape, the outputs of the first four recorded tracks were returned to the console for mixdown to two tracks. These outputs were then "bounced" to the two open tracks, 7 and 8. This transfer was accomplished with tracks 1 through 4 in the sync reproduce mode (newer ATR systems provide high-quality sync output, allowing repeated bounces to be made with little signal degradation or noise buildup).

With careful planning, it is possible to create a large-scale production on a limited number of available tracks.

Fig. 5-3. Track bouncing is a production technique used in multitrack recording to expand the number of available tracks and thus allow for additional overdubs.

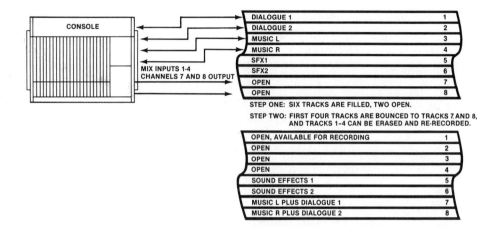

Phases of Audio-for-Video Postproduction

The process of postproduction as a whole is one of manipulation of original source material to create a final program; as such, it extends from the end of the production stage to the final edited video master tape.

Postproduction of audio for video may be divided into five steps, or phases: spotting, laydown, sweetening, mixdown, and layback.

Spotting

The first step in audio-for-video postproduction is that of *spotting* the soundtrack for additional sound cues beyond those existing within the original sound audio. Spotting is the process of planning out each of the many elements that go into the video soundtrack. This includes such aspects as making arrangements for the logical and effective placement of dialogue, sound effects, and music within the overall soundtrack.

Depending on the scale of the production, a spotting session may involve several specialized sound editors, each dealing with a specific aspect of the overall sound script. However, for smaller productions, the director or producer may perform all the necessary tasks. The spotting session in either case requires the use of a dub (generally recorded in the 3/4-inch U-matic format) of the final edited master, with time code burned into the visible picture or with VITC and a character inserter for visible time code at all tape speeds, including still frame.

During the actual process of spotting, the sound engineer or editor reviews the dub, whose time code addresses are identical to those of the edited master, noting in and out points for the various aspects of the sound design. Since the time code may be read at any forward or reverse shuttle speed, including still frame, a spotting editor is able to locate individual cue points with frame accuracy.

Production decisions such as the identification of potential trouble spots are included within this process. The spotting editor may, at any point, go back and review the original source material in order to determine the quality of the recorded dialogue or sound effects. Through careful review, potential trouble spots may be given advance consideration and corrected. For some problems, however, the solution may call for the rerecording of the sound during postproduction.

The final product of this phase is an audio cue sheet in the form of either a sound cue edit decision list for on-line postproduction or a reasonable facsimile for off-line postproduction.

Laydown

In certain nonmusical and EFP work, the multitrack ATR may be brought into audio-for-video postproduction in a phase known as *laydown*. Laydown is the process of transferring synchronized source audio tracks of a video program to a multitrack tape via an ATR for postproduction purposes. At this point, audio and video are effectively divorced.

The source audio encountered in laydown may originate from any combination of the following:

- Original single-system audio from field videotapes
- Original double-system audio field tapes
- Edited audio from the edited master videotape

Source dialogue, original sound effects, and sound presence may be transferred onto a consolidated multitrack format for further creative postproduction manipulation.

Methods of Laydown
Like the video edit process, the laydown process may be carried out with the aid of an on-line edit controller or through the use of basic off-line synchronization techniques.

On-Line Laydown—With the on-line method of laydown (Fig. 5-4), an EDL-based system is used to define edits. These edits are specific sections of audio that are to be copied from a source audio track onto a multitrack ATR. An edit screen on a CRT may be used as a quick reference on which to check the status of the system, the time code locations of the master, record, and source machines, and information about the current edit and other edits that may have already been completed.

Fig. 5-4. System for on-line laydown.

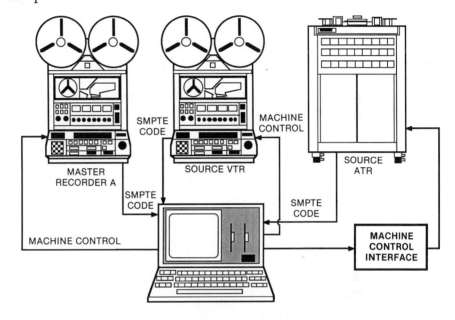

In essence, the laying down of an edit under computer control involves obtaining the time code addresses of the in and out points on the master, thus giving indications as to where the edit is to be laid down. Similar time codes are entered into the system for the source tape's in and out points, yielding a time code "map" for the precise placement of audio segments onto the master multitrack tape. Once entered, time code addresses may be compiled into an edit decision list for later automated assembly.

When an actual edit is to be made, the operator invokes the record command, and the system will automatically cue the required machines to the correct positions. The controller will then switch the transports into the play and chase mode until the edit is accomplished, that is, until the source material has been copied onto the multitrack tape in the proper location.

Off-Line Laydown—The laydown process may also be carried out in an off-line fashion, utilizing a time code synchronizer (Fig. 5-5) instead of a video-type edit controller.

Fig. 5-5. System
for off-line
laydown.

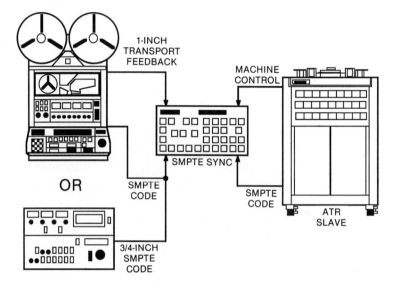

In off-line laydown, the audio source material may be derived from single-system or double-system field masters. However, the general approach is to transfer the edited audio tracks and the time code from the final edited master directly onto the multitrack tape of an ATR. Using this approach, the audio tracks of the edited master, which were built up in the video edit, become either a reference scratch track or the finished audio track for the video program. In the latter case, extra care must be taken during the video editing process to end up with clean, properly mixed tracks, since these tracks will largely determine the overall audio quality of the finished program.

Time Code in the Laydown Process

Since the laydown process is based on the precise location of synchronous audio material, accurate time code must be present on all source and master tapes.

In practice, the means by which time code is laid down on the multitrack master tape differs in the on-line and off-line methods. Using the on-line method, the multitrack tape of the record transport must be prestripped with time code, and the source machines are synchronized to it and transfers are made using time code offset calculations (match-frame techniques), which are performed by the edit controller. Off-line laydown often requires the audio tracks and time code of the video edited master to be directly transferred to the multitrack ATR for further work within postproduction. Whenever possible, time code identical to that of the edited master should be used in off-line production, thus avoiding potential confusion due to the manual offset calculations that must be made.

It is important to note that, in the transferring of time code to a multitrack ATR (or any other ATR or VTR), the time code must be regenerated using the jam-sync output in order to avoid degradation of its signal.

Sweetening

Sweetening refers to the addition of musical or nonmusical elements to an existing audio track in order to heighten the impact of the complementary visual material. With respect to audio for video, the process of sweetening may refer to a broad range of postproduction techniques. The chosen course of action is entirely dependent on the individual requirements of the program and may range in scope from the single layering of a background presence track to the enhancement of an important visual cue with full production dialogue, presence, sound effects (SFX), and orchestral scoring. The necessity for synchronized multitrack audio techniques for the more involved production ventures is evident. The ability to store isolated sound cues and yet maintain a synchronous time relationship among them serves to make the process of sweetening feasible, regardless of the storage medium (multitrack ATR or computer-based digital ATR).

Time Code in the Sweetening Process
Within the postproduction phase of sweetening, time code continues to serve a central function. Time code is generally required for the addition of synchronous sound cues to the multitrack tape or for the synchronizing of electronic instrumentation and sound effects through a SMPTE/MIDI lock. However, one of the greatest assets of using time code in the sweetening process is the ability to maintain control over tape location and basic transport functions with frame accuracy and repeatability.

If time code is utilized in this phase, the time spent earlier on the spotting of dialogue, music, and SFX more than pays for itself. The time code location of a desired sound cue may be easily entered into the synchronizer or edit controller, which will then chase the source or master transport to the preselected location for execution of the transfer to the multitrack ATR.

Track Building

Within the video world, the basic concept of sweetening may be encountered as part of the process known as *track building*. Track building may be viewed as the judicious transfer of synchronous or nonsynchronous audio to a multitrack ATR in order to produce a final video soundtrack.

The first stages of track building are often encountered during the video edit, at the point when the contents of the original source videotapes are transferred to the final edited master in an on-line fashion. Dialogue, narration, and/or sound presence can be recorded on the edited master under the control of an EDL-based video controller.

The building of audio tracks during the video edit is a necessary phase of the production of a video project. Such tracks may take any of three possible forms:

1. Reference scratch track for the spotting phase and for the buildup of audio within the sweetening phase

2. Basic dialogue and background track to which sweetening is to be done

3. Finished audio program track

A larger video production may employ all of these forms (depending on the scope and requirements of the specific production); however, projects operating with limited equipment, time, or money may be forced to stick to either of the latter two options.

In this phase of postproduction, where original audio is to become a part of the final soundtrack, it is wise to pay attention to maintaining a clean set of tracks. This is accomplished by careful adjustments of level balances and the matching of tonality on both dialogue and presence tracks. This process may be made easier by the fact that most professional VTRs provide two audio tracks in addition to a dedicated time code track. This allows the engineer to isolate the dialogue and presence or SFX tracks throughout the entire process of track building.

After the sound from the final edited videotape and/or the original single-system or double-system audio tape has been laid back to the multitrack ATR, the basic audio program is ready for the building of additional tracks via the process of sweetening. The plan used to map this process is the audio script that was compiled during the spotting phase.

Once the required audio material for further postproduction has been gathered, the multitrack master and a time-encoded window dub of the edited master videotape may be brought together in a facility known as the *sweetening bay* (Fig. 5-6). The sweetening bay is designed to facilitate the transfer of prerecorded audio onto a multitrack ATR in order to enhance the audio portion of a synchronized video program. The equipment found within the bay is of a number of formats, providing for both synchronous audio (time-encoded ATRs, event-triggerable sampling devices, and event-triggerable, programmable compact disc sound effects) and nonsynchronous (also known as wild) audio (ATRs, phono turntables, and broadcast-style cartridge machines).

The audio material that is to be worked with in the sweetening bay may originate from various sources, for example:

- Original and synchronous dialogue, narration, or effects
- Original and nonsynchronous dialogue, narration, or effects
- Originally recorded music scoring
- SFX or music library on records or compact discs
- Personal (recorded) SFX library

Within the track-building stage of sweetening, dialogue, narration, presence, and SFX are laid onto the multitrack ATR in accordance with the overall plan contained in the audio spotting sheet.

The Sweetening Process in the Recording Studio

Even though the professional recording studio has been the main source of music scoring and dialogue production for the film and video industries for many years, recently there has been a proliferation of audio-for-video recording facilities. This has occurred as a result of expectations of an increased revenue base in the video industry, which have been fulfilled to a great extent. Recognizing the economic gains that may be realized from new production strategies, many recording studios have added such equipment as video monitors and synchronizers to their rosters. However, the production of audio for video requires more knowledge concerning (even a degree of specialization in) the control of time code devices and signal-processing equipment than is evident at first glance. Through recent improvements in control synchronizers and audio edit controllers, control over ATRs, VTRs, and event-based programmable devices has finally been brought within the economic and technological range of the recording facility that is trying to break into the competitive video market.

The professional recording facility's role in the production of audio for video may be quite diverse and may include such formats as automatic dialogue replacement (ADR), sound effects (SFX), electronic music, and live music scoring and mixing.

**Fig. 5-6. The
sweetening bay
is used in audio-
for-video
postproduction.**
*(Courtesy of Sony
Corporation of
America, Inc.)*

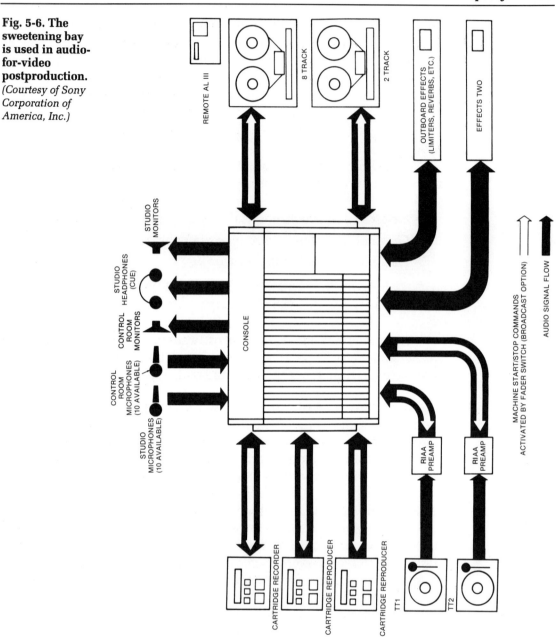

Automatic Dialogue Replacement

Often the conditions under which a project's original dialogue or sound presence is recorded are less than ideal. As a result, master field tapes are brought into the video edit or laydown process with an inadequate or unusable set of audio tracks. When such a thing happens in a larger scale production, rather than reshoot the original scene, it is generally preferable to *rebuild* an audio track within a suitably equipped recording studio or sound stage. In fact, under certain conditions, a production team doing EFP may opt to record a set of scratch audio tracks, which are then replaced within the more controlled atmosphere of a recording studio.

In the rebuilding of audio during postproduction, the main concern is most often with on-camera and off-camera dialogue. This is because most of the presence and SFX tracks are usually added later, during the general process of sweetening. The technique used to rebuild a track containing dialogue is called *automatic dialogue replacement (ADR)*, also commonly known throughout the industry as *looping*. The latter term is taken from the name of an equivalent technique long used in the motion picture industry for the rebuilding of audio soundtracks.

The process of film looping involves the creation of three continuous loops of film, each of precisely the same length. The first loop is a rough cut of the scene for which the audio is to be rebuilt. The remaining two loops are a mag-track (magnetically stripped film for the recording of audio) dub of the original audio and a clean portion of mag film for the recording of fresh audio. These continuous loops are loaded onto three transports, and they may then be viewed and monitored within the studio in continuous and synchronized fashion, allowing the actors who are to be recorded to practice their timing. The actors will monitor the original scratch track and the newly recorded dialogue over headphones, until they have mastered their parts. Finally, the dubbing engineer will switch off the original audio (to allow the artists to concentrate better and to reduce the possibility of leakage onto the replaced track) and the rebuilt audio will be recorded onto the fresh mag track for later resynchronization with the final edited film. This process is continued until all required tracks have been properly rebuilt.

Automatic dialogue replacement, whether for film or for audio production for video, places great demands on both the performing artist and the sound engineer. It is often difficult to duplicate the on-camera dialogue exactly with respect to both timing and perspective. The repetition of each scene that is to be replaced enables the artist to practice until a satisfactory replacement track has been achieved. The sound engineer must also be attuned to the requirements of each scene. For example, an on-camera shot showing a performer walking toward the camera will require careful matching for the sound to have the correct level or sense of perspective.

The sound stage or studio that is used for ADR should be rather large and exhibit slightly "dead" acoustics. In such an environment, the microphone can be placed at greater distances from the source when that is called for to match the perspective in the visual scene.

In audio-for-video production, the techniques of looping are much the same as those used in film. However, the use of control synchronizers and edit controllers whose functioning is based on time code has eliminated the need to create separate loops, thus allowing ADR to be accomplished more quickly and easily. Many of the newer control synchronizers found on the market are equipped with the ability to create a loop electronically, through the use of control and synchronization techniques based on time code. The time code address representing the first frame of the loop is entered into the controller, along with the address of the end frame. Entering the looping mode, the controller initiates a play-recue-play sequence, replaying the desired section of film as many times as necessary, until the dialogue has been replaced by switching into the record mode for the proper audio track.

A device that was specifically designed for the purpose of automatic dialogue replacement is the Taker A/B (Fig. 5-7), which is manufactured by Giese Electronic of Hamburg, Germany. The Taker A/B consists of a remote control keyboard (Fig. 5-8) and a 19-inch, rack-mountable central processor unit. This device does not operate as a controlling device over multiple transports in the sense that a control synchronizer does. Instead, an internal time code reader is used to determine program location on the time-encoded dub of the edited video master. The device is able to maintain control over the videotape location and to loop the program material by means of in and out points. Audio for automatic dialogue replacement may be recorded in several ways, such as onto the audio tracks of the videotape itself or onto open tracks of tape on a synchronized ATR that has time code identical to that of the edited video master.

Through the use of the Taker A/B, satisfactory audio for ADR may be recorded onto track A of the VTR or synchronized ATR, and then an attempt at a better track may also be recorded onto track B, allowing the first take to be kept as a backup.

Fig. 5-7. The Taker A/B automatic dialogue replacement controller. *(Courtesy of ESL, Inc.)*

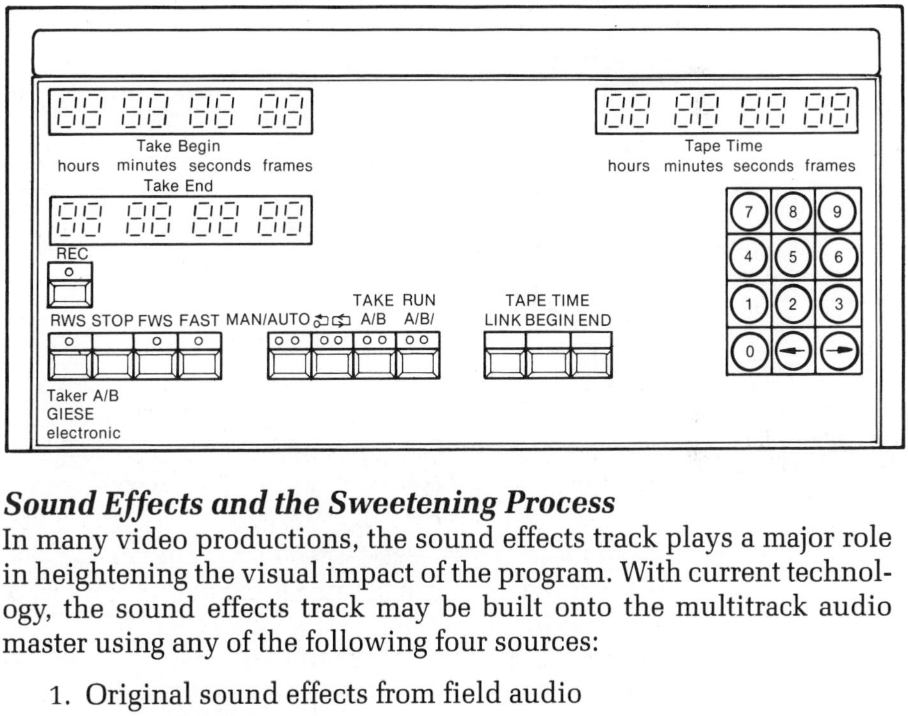

Fig. 5-8. The Taker A/B remote control keyboard. *(Courtesy of ESL, Inc.)*

Sound Effects and the Sweetening Process

In many video productions, the sound effects track plays a major role in heightening the visual impact of the program. With current technology, the sound effects track may be built onto the multitrack audio master using any of the following four sources:

1. Original sound effects from field audio
2. The Foley sound stage
3. Sound effects libraries
4. Electronically synthesized sound

Original Sound Effects from Field Audio—For some video shoots on sound stages or on location (EFP), the most effective and desirable sound effects that may be captured on tape are the original on- or off-camera effects. If original sound effects are to be utilized, it is the goal of the sound recordist to capture that sound on tape as cleanly as possible. Such a track may be recorded in the field onto the second track of a single-system VTR or onto a separate track of a double-system ATR. Once brought into the postproduction phase, the sound effects may be rerecorded onto the edited video master or directly onto a multitrack ATR, depending on the production approach chosen.

The Foley Sound Stage—In the early 1940s, George Foley, then a sound mixer at Warner Brothers, designed a process by which on-camera ambient sounds that had been lost in doing ADR could be replaced. This process, known as the *Foley process*, is somewhat similar to ADR in that the picture and audio are run in an electronic loop to allow for artists' practice and then replacement of the sound. All necessary props that are to be utilized in the synchronized replacement of on- and off-camera sounds (for example, doors slamming, footsteps, clothes rustling, doorbells ringing, etc.) are gathered and then recorded onto separate audio tracks within the sound studio.

The miking techniques involved in this process are similar to those for ADR, and a large studio with slightly "dead" acoustics is preferred in order to maintain the proper distances to achieve the perspectives required by the individual scenes. Additionally, a Foley sound stage is often equipped with a variable-surface floor (Fig. 5-9), allowing for the recreation of the sound of footsteps on specific types of surfaces, such as hardwood, gravel, and cement.

Fig. 5-9. The variable-surface floor found in a Foley sound stage.

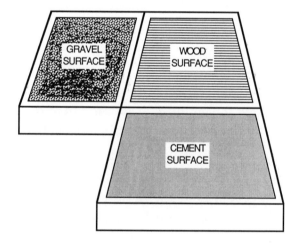

Sound Effects Libraries—In addition to original field audio, another major source of sound effects is the sound effects library. The sound effects library offers a wide array of easily accessible sounds for use within the sweetening process for broadcast or production audio. A number of these libraries are currently available; each contains recorded effects ranging from various types of background noise, gun shots, and cars crashing, all the way to the far-off strains of a bagpipe. Such libraries do have differing degrees of comprehensiveness. The complete library offered by Network Production Music of San Diego, California, contains more than 12,000 sound effects recorded on 87 albums, with an accompanying indexed catalog of approximately 480 pages. However, smaller libraries that offer less diversity may be sufficient for a small-scale production.

When sound effects that are recorded on a record album are to be recorded onto the multitrack tape containing the synchronized program sound, it is often necessary to first transfer the desired effect to a medium that can be more easily cued or even one that can be synchronized to the edited master. For example, sound effects from a record album may be transferred to:

- 1/4-inch nonsynchronous audio tape (for *hot-rolling* a nonsynchronous effect by way of preroll cueing onto an assigned track of a multitrack ATR)
- Time-coded synchronous audio tape (for precision timing of the recording of an effect on a multitrack ATR, utilizing LTC frame-offset adjustments within the track-building process)
- Time-encoded synchronous videotape (for precision timing of the recording of an effect on a multitrack ATR, utilizing LTC or VITC frame-offset adjustments within the video edit process)

Sound effects albums are generally available for production use as one-time buy-outs at $10.00 to $20.00 per album, with no additional charges or per-use licensing fees.

Although sound effects libraries most often take the form of standard record albums, many libraries are beginning to appear in the more robust format of compact discs (Fig. 5-10). Recent advances in the storage and retrieval of sound effects and music using the compact disc format have made this medium very desirable for many audio postproduction and broadcast applications. In addition to superior audio quality, compact discs offer additional features to the engineer working in audio-for-video production:

- Event triggering means that precision cueing to a time-encoded production device is possible without the necessity for internal time code; this provides a high degree of accuracy and repeatability.
- Since compact discs may be easily and inaudibly looped, they can be used in the laydown of background sound presence.

- Since compact discs are computer-based, they may be accessed and/or cued by means of a portable computer.
- With a computer-based index, the SFX library on a compact disc may be indexed and cross-referenced within a data base for ease of accessibility.

Fig. 5-10. A sound effects library on compact disc. *(Courtesy of Valentino, Inc.)*

Although major breakthroughs in compact disc technology may be expected within the next few years, systems such as the Sony CDP/CDS-3000 compact disc player (Fig. 5-11) are already being utilized in current production. The Sony system, which consists of a central controller unit and one or two compact disc players, was designed to provide easy access to audio content as well as programmability. Both player units may be programmed at the central controller to operate manually or to play up to eight selections automatically, in any order requested. Once the program play mode has been initiated, additional selections may be loaded into memory as openings become available. In the manual or monitor play mode, any point on the disc may be directly accessed via the 10-key pad, with an accuracy of 1 frame or 13.3 milliseconds. A search dial allows the user to perform a manual bidirectional search, while listening to the reproduced sound. If the player is in the manual mode, a full 360° turn of this dial corresponds to a 1-second shift (75 frames) on the disc, either forward or backward, depending on the direction in which the dial is turned. If the dial is rotated when the player is in the monitor play mode, the shift is from 20 to 60 seconds, depending on the speed at which the dial is rotated.

Fig. 5-11. The Sony CD Modular System.
(Courtesy of Sony Corporation of America, Inc.)

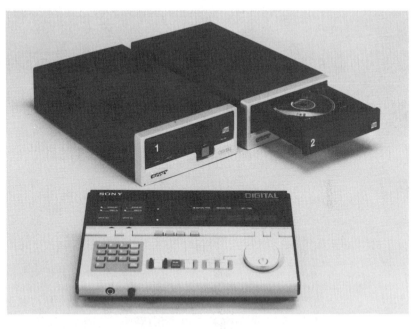

Music Scoring and the Sweetening Process

Generally speaking, the techniques employed in music scoring for audio production for video are similar to those used in standard record production. The basic difference is that in audio-for-video production there is a need for systemwide ATR and VTR synchronization and transport control and precision timing between the audio and visual aspects of the program. The requirements placed on the audio-for-video production facility differ from those affecting the music recording studio because the musical material to be recorded for video is designed to exist in conjunction with the visual aspects of a program. The composition of a music score for a video scene may be conceived as a foreground driving force, propelling a scene forward, or it may be written in order to suggest or intensify a certain mood. In either case, the addition of effective music scoring to a scene almost always serves as a valuable asset to a video production.

The art of music scoring for video or film attempts to create a direct interaction between sound and sight. From the very early stages of preplanning through the production phases of a major project, the composer of the music score will often be involved in an interactive selection process. He or she will select the type of music, the orchestration, and the tempo so as to match the mood and foster the degree of involvement required by each scene.

The process of scoring is based on the concept of precisely matching the musical structure of a composition to the visual aspects of an edited video worktape. Often the timing constraints on the scoring process are so tight that an extremely steady tempo must be maintained in order to precisely time the passage of the musical measures (a process often more closely related to the mathematics of video frames than to the art of music composition). A device may be used to generate a *click track*, that is, to deliver a precision metronomic beat to a spare track of the multitrack ATR and also to a headphone monitor. In this way, accurately timed musical measures, whose lengths are precisely divisible by the lengths of video or film frames, are created, allowing for accurate editing and arrangement of the musical content with respect to the visual action of a program.

Once the score has been shaped into its final (or roughly final) form, the music director for the recording session (who may or may not be the composer) undertakes the task of organizing and conducting the studio musicians in a performance of the score. The musicians perform the music segment by segment to a playback of the edited video worktape. Meanwhile, equivalent (jam-sync) time code is laid down on the multitrack ATR so that synchronization and layback to the final edited master videotape can be done later.

Prerecorded Music Libraries—Original music scoring brings the greatest degree of flexibility and impact to a production through the creation of a piece of music to match a visual scene. However, this option may not be cost-effective or feasible for a given production. In such a case, the producer may opt to make use of a prerecorded music library.

Like sound effects libraries, music libraries are most often packaged as record albums, with compact discs becoming increasingly popular.

Currently, there are more than 25 sources for music libraries for use in the audio sweetening bay. The available libraries vary widely as to the extent or scope of the music they contain. The companies that market the libraries generally offer a customer service that allows users to request and audition a specific sound cue. This improves a producer's chances of getting the sound that best suits the needs of a specific production.

The fee structure encountered when using a music library may involve any of several possible methods of licensing. The most commonly found is for the producer of the music library to receive a usage royalty, or a *needle-drop fee*. This means that every time a specific music cut is used within a production, a charge is applied. If two needle drops are made in order to lengthen a sound cue, a double fee will be charged to the user. Music library companies may also base their charges on the number of running minutes used within a production. Others offer the option of paying a one-time charge for the use of a library for a one-year period. The latter arrangement greatly reduces the paperwork necessary for repeated production work.

Typically, a postproduction sweetening facility will have one or two complete libraries available for reference, and records of individual sources may be kept on hand for specific applications. The creation of a customized index for the available music libraries may be a difficult task, but it is one that is necessary. An index may take the form of a bound folder containing listings of material, with comments and updates, or it may be maintained within a standard or customized computer data base, allowing for the cross-referencing of selections.

Electronic Music and the Sweetening Process

During the past decade, the electronic music synthesizer and other electronic music devices have gone from being regarded as mere curiosities to becoming almost irreplaceable devices for music scoring and sound effects production in the media of film, video, and music recording. Their high degree of flexibility and cost-effectiveness has made synthesizers standard equipment within the audio-for-video production facility.

Electronic music devices may be divided into four general categories: analog music synthesizers, digital music synthesizers, pitch-controllable sampling systems, and event-triggered sampling systems.

Analog Music Synthesizers—Analog music synthesizers (found as either keyboard devices or instrument-controlled devices) are used to generate music and/or sound effects and incorporate audio oscillators, waveform filters, VCAs (voltage-controlled amplifiers), and VCOs (voltage-controlled oscillators). These synthesizers may be programmable and are essentially hybrid analog units that are controlled digitally for ease of operation and repeatability of programming.

Digital Music Synthesizers—The digital music synthesizer generates music and/or sound effects by employing digital algorithms that closely approximate the equivalent analog sound. These microprocessor-based devices produce polyphonic digital sound that may be closely controlled and reconfigured for specialized production applications. The latest models of digital synthesizers often incorporate computer-style keyboards, CRT screens as monitors, and even light pens that can be used to draw a waveform directly onto the screen. Unlike its analog counterpart, which is limited to a "building block" approach by its finite number of oscillators and filters, the digital synthesizer is able to create an almost infinitely varied number of sounds or "voices," with absolutely accurate recall.

Pitch-Controllable Sampling Systems—A sampling system is a device that is capable of capturing a specific natural sound and digitizing it for further processing and reproduction on demand. A pitch-controllable sampling system often includes an instrument with a musical keyboard. Examples are Fairlight C.M.I. Series III, Synclavier 2, Kurzweil 250, Emulator 2, Prophet 2000, Ensoniq Mirage, and recently introduced units from Roland and Korg.

A top-of-the-line pitch-controllable sampling system is the Fairlight Series III (Fig. 5-12). This device includes hardware for the reproduction and resynthesis of up to 16 voices from an original sound signal. It also offers 8 megabytes of waveform memory (expandable to 24 megabytes, representing more than 2.5 minutes), 16-bit stereo A/D conversion at a sample rate as high as 50 kHz, and floppy and 60-megabyte hard disk drives. Ports for inputting other electronic instruments and SMPTE time code are provided, in addition to an alphanumeric keyboard with a touch-sensitive graphics tablet, a 6-octave touch-sensitive musical keyboard, a system controller, and a video monitor. Software for the Series III system is available to perform sampling,

Fig. 5-12. Fairlight Series III system. *(Courtesy of Fairlight Instruments.)*

waveform drawing and synthesis, two forms of sequencing, music printing, Fourier analysis and resynthesis, and telecommunications.

Event-Triggered Sampling Systems—Event-triggered sampling systems operate similarly to other sampling devices but are intended to reproduce the original signal in real time or pitch-shifted time and are not capable of polyphonic resynthesis of that signal. This type of device is, in effect, a random-access reproducer of short-duration musical and sound effects. Modern studio drum machines, triggerable digital delay lines (DDLs), and digital signal processors (DSPs) fall under this heading.

Musical Instrument Digital Interface

Within present-day music recording facilities, a large array of electronic instruments may be used simultaneously for a single production. In order to provide a synchronization and communication linkup of devices such as drum machines, synthesizers, emulators, and sequencers, it is necessary to utilize communications software called *musical instrument digital interface*, better known as *MIDI*. MIDI-based communication between digital electronic instruments is generally necessary in the synchronization and distribution of musical information during electronic music scoring.

The software language that is included in the MIDI structure is designed to transmit two forms of musical data: (1) a specialized language for describing and controlling musical events in real time, and (2) a "channel" specifying the distribution of information to a particular track or piece of equipment.

The MIDI system processes blocks of machine-language data known as *words* that describe or request specific musical events. The communicated words for the description (send information) and request (receive information) of a specific event are identical in structure.

The operation of MIDI is such that if a note (for example, middle C) is played on a MIDI-equipped synthesizer, the instrument will sound the required note and will also transmit an encoded *phrase* as output at its MIDI port. Such a phrase may be read as follows: [Note Event: ON, Pitch: Middle C, Velocity: (Release Velocity Value)]. If this MIDI phrase is recorded by an internal or external sequencer, the same phrase structure can be precisely reproduced in perfect synchronization with any simultaneous musical data. That is, the synthesizer will again play the previous sequence even though no note has been physically touched on its keyboard.

In addition to creating the basic word structures, MIDI controls intelligent distribution of the words to specific destinations, even when serial connectors are joined in a "daisy-chain" fashion (Fig. 5-13). The designation of a device destination within the MIDI-encoded word causes information to be sent to the correct device among several that are connected by a single communications linkup, or chain. This concept of channel distribution allows a MIDI controller (a portable computer or similar device) that has one interface port to communicate with multiple electronic musical instruments of similar or dissimilar function (drum machines, synthesizers, sequencers, etc.), multiple voice and/or split keyboard devices, or any combination of the preceding, with a maximum single-line distribution of up to 16 channels.

Fig. 5-13. Electronic musical instruments communicate using MIDI interconnections in a "daisy-chain" fashion.

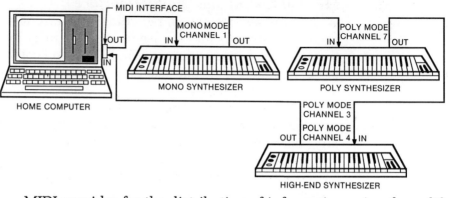

MIDI provides for the distribution of information using three different communication modes (Fig. 5-14), which may all be easily reconfigured at any time:

- In the *omni mode*, a music device will respond to any set of words that is distributed along the information chain, regardless of the source.
- In the *poly mode*, a music device will only respond to words it receives from an assigned set of channel sources.
- In the *mono mode*, a music device will only respond to words sent from one specific channel.

Utilizing this distribution hierarchy, the controller and the controlled devices can maintain musical synchronization at all times, without the need for external control devices.

The hardware used to provide the communications link between devices is a 5-pin DIN (European specifications) connector. The connected devices are electro-optically isolated, in order to prevent ground loops and to avoid unnecessary pickup of noise. The encoded data are transmitted throughout the system at a serial baud rate of

Fig. 5-14. MIDI communications modes.

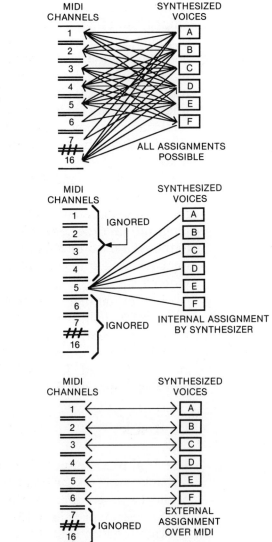

(A) The omni mode.

(B) The poly mode.

(C) The mono mode.

31.25K baud and, therefore, less than 1 millisecond is required to transmit note and velocity data.

Mixdown

When all the desired audio segments have been recorded onto a multi-track ATR and have been sweetened to the satisfaction of the sound script or score producer, the next phase of postproduction is *mixdown*, or the *remix phase*. The purpose of this phase is to creatively blend the separate tracks existing on one or more synchronized multitrack machines into a final balanced soundtrack, which may be in stereophonic or monophonic form. This task is accomplished by putting the status command at the audio production console into the mixdown mode (or by placing the microphone/line switches for the appropriate console inputs into the line position). In this way, the line outputs of the multitrack ATR(s) are fed to their respective inputs at the console. The multitrack tape is then repeatedly reviewed while adjustments to level, equalization, effects, reverberation, and panning are made on each track.

Once the final music and/or sound blend has been approved by the producer, the audio material, which is thereafter known as the *final mix*, is rerecorded on an ATR in stereo or mono, along with equivalent longitudinal time code (Fig. 5-15). Jam-synced LTC must be laid down to the ATR used for mixdown in order to allow for the final stage, laying back the audio content of the program to the 1-inch edited master videotape.

The format that is currently most popular for the purpose of mixdown is 1/2-inch, 4-track ATR operating at either 15 or 30 ips. However, 1/4-inch half track with a center track for time code, a digital ATR, or a digital processor incorporated with a time-encoded synchronized VCR may also be used. Track standards for these formats are given in Section V of Appendix B.

Layback

The final phase of postproduction of audio for video is *layback*. The layback process is the point at which the separated audio and video aspects of a program are rejoined in synchronization on a 1-inch edited master videotape (Fig. 5-16).

Tight control over the synchronization of VTR and ATR is necessary within this phase. Frame offset adjustments may be necessary in order to bring program aspects into absolute sync. Throughout all postproduction phases of audio for video, when a VTR is used as the layback machine, it is necessary to designate the VTR as the master transport for the purpose of time code provision, and the ATR must be designated as the slave machine.

The process of laying back the audio program onto the edited master videotape may be carried out via several distinct options (Fig. 5-17):

1. *Layback directly to the VTR.* The mixed-down version of the audio program, containing LTC, is synchronously transferred

Fig. 5-15. Typical audio-for-video mixdown configuration within a recording studio. *(Courtesy of Sony Corporation of America, Inc.)*

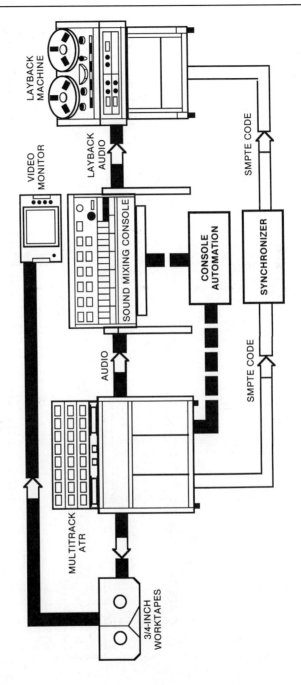

Fig. 5-16. Audio layback in an audio production facility.
(Courtesy of Sony Corporation of America, Inc.)

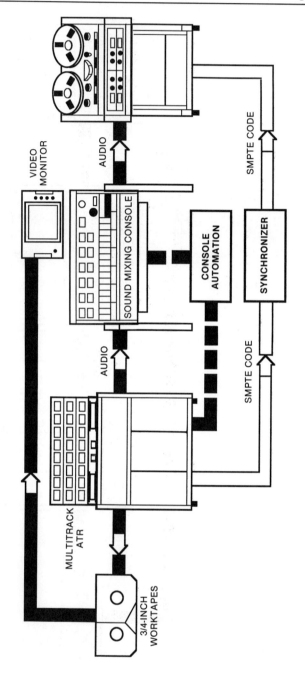

to the 1-inch edited master videotape. This is accomplished by copying a synchronous dub of the mixed-down soundtrack to the 1-inch Type-C VTR, which has been placed in the audio record only mode. In this way, final audio is recorded on the same tape as the video program material for distribution or broadcast.

2. *Layback to a layback ATR.* In recent years, several tape transports have come on the market whose electronics and head configuration are designed to match precisely with the audio portion of the 1-inch Type-C video tape recorder. These devices are known as *layback ATRs* (Figs. 5-18 and 5-19). By using a layback ATR, it is possible to gain such production advantages as improvement of audio quality as a result of optimized alignment to the videotape in use and performance of audio layback without having to tie up the much more expensive and versatile Type-C VTR.

3. *Layback directly to a layback ATR from mixdown.* Through the use of a synchronized layback ATR, it is possible to transfer the audio material from the multitrack ATR(s) directly to the audio tracks of the edited master videotape. By doing this, the layback transfer process may be avoided, circumventing the associated audio loss of a generation. Unless the audio program material is very simple in nature, it is wise to perform this process with a console equipped for signal level automation. Once a final mix has been built and approved, the direct layback process is simply a matter of prerolling the edited master videotape onto the layback machine and initiating the record commands.

The Music Recording Facility

During the 1970s, the concept of synchronization was introduced in music recording studios as a means of increasing the number of tracks available to record producers and artists (through the synchronization of two multitrack recorders). In the 1980s, the recording studio has been exposed to a steadily increasing flow of multimedia production and control techniques that rely on time code synchronization. The great number of audio-for-video control devices presently being marketed to the audio recording industry means that it is more important than ever for professionals in the video and broadcasting industries to

Fig. 5-17. Audio layback in a video production facility. *(Courtesy of Sony Corporation of America, Inc.)*

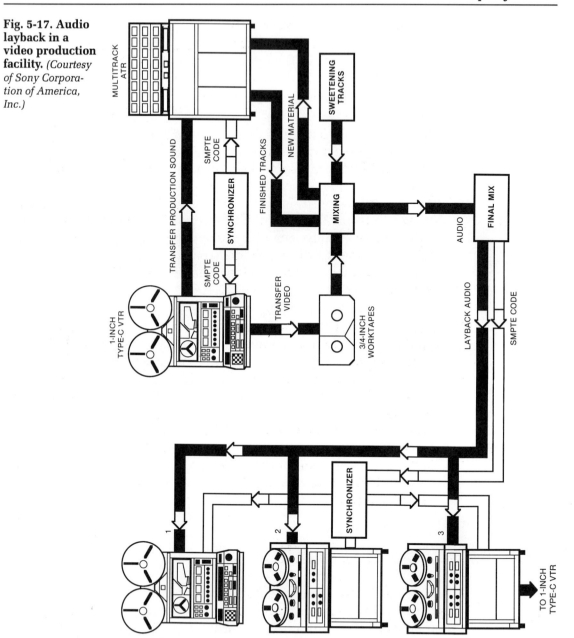

have an understanding and, in certain instances, a working knowledge of the music and sound recording process.

Fig. 5-18. The Sony JH-110B-LB-VP audio layback recorder reproducer. *(Courtesy of Sony Corporation of America, Inc.)*

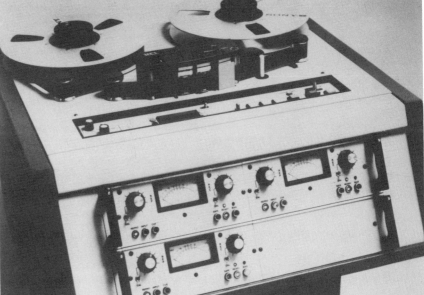

The modern multitrack recording facility (Fig. 5-20) is multifaceted. To function, it draws on such diverse fields as music, acoustics, electronics, music production, audio-for-video production, business, and (ask any recording engineer) psychology. All of these elements are blended together to create a single end product—a final master recording or soundtrack.

As well as having many facets, the process of recording also involves many stages in order to produce a master recording or soundtrack. Preproduction planning, recording, overdubbing, and mixing are some of the stages in this process.

Personnel in the Recording Facility

In order to understand the environment and operating practices of the recording facility, it may be useful to know something about the roles of the personnel who are involved in a music or audio-for-video recording session.

The Engineer

The engineer's job can perhaps best be described as that of "an interpreter in a techno-artistic field." It is his or her task to give substance to

Fig. 5-19. The Studer A80VU-3 LB audio layback recorder. *(Courtesy of Studer/Revox of America, Inc.)*

the artist's creation through the medium of recording. Accomplishing this is really an art, since the technologies of music, recording, and audio-for-video production are somewhat subjective in nature, reflecting the tastes and experiences of those involved.

During a recording session, a free-lance or house engineer will generally place the musicians in the best locations, choose and place the microphones, set levels and balances on the recording console, and record the music onto tape. During an overdubbing, mixdown, or layback session, the engineer will use the console to create a balance from the multitrack machine for recording or mixdown purposes.

The Sound Mixer
The task of the sound mixer is somewhat like that of the studio engineer; however, it is strictly limited to providing the final balances within the mixdown phase. Often the sound mixer is called on to deal

effectively with the overall balancing of dialogue and effects, in addition to the final music balances of the program.

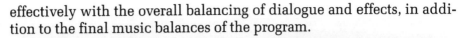

Fig. 5-20. A control room in Criteria Recording Studios. *(Courtesy of Criteria Recording Studios; photograph by Carlos Domenech.)*

The Assistant Engineer

Larger recording facilities sometimes train future staff engineers by encouraging them to work as assistants to the house engineer. The duties of an assistant engineer may include setup of microphones and headphones, operation of and control over tape machines, breakdown of equipment at the end of a session, and, in certain cases, adjustment for a rough mix of balances at the console for the engineer prior to an overdubbing session.

In the past, particularly in English recording facilities, seconds (as assistant engineers are also known) were instrumental in the smooth functioning necessary for complex recording sessions. However, with recent innovations in control and synchronization technology, the involvement of the assistant engineer has seen a minor decline.

The Maintenance Engineer

It is the job of the maintenance engineer to see that equipment in the studio and control room is properly maintained and aligned, with the correct equalization and record level values, so that it will operate efficiently.

Larger organizations, which operate more than one recording facility, often employ a full-time maintenance engineer on the staff. Smaller studios or ones located far away from other facilities within their organization may hire the services of a free-lance maintenance engineer on an on-call basis. Often, when no maintenance engineer is on the staff, the studio's engineer will maintain and align the equipment.

The Producer

The audio producer is often hired by the artist or record company on a free-lance basis. He or she will generally oversee and direct the recording process, giving personal opinions as to how to make the project turn out as well as possible.

With respect to the recording of a soundtrack for video, it is important to keep in mind that not all music producers are familiar with the synchronization and control techniques required to produce quality audio for video. A recording facility that has a staff that is experienced and qualified in audio-for-video production will be a great asset to the producer who is new to this recent and popular subdivision of the music industry.

Physical Layout of the Recording Facility

The professional recording facility actually consists of two distinct working areas: the recording studio and the control room.

The Recording Studio

The *recording studio* is a room or series of rooms that is acoustically tuned or optimized for the purpose of placing the best sound possible onto a tape track, when a microphone pickup is used. The studio is an acoustically isolated structure, which is designed to keep outside noises from entering and thus being recorded on tape, while also keeping internal sounds from leaking out and disturbing the surrounding neighborhood.

Studios vary in size, shape, and acoustic design according to the specialized music styles preferred by and the personal tastes of the owners and staff. For example, a studio in which a great deal of rock and roll is recorded may be small in size, with a highly absorbent acoustical treatment to allow for high volume levels and separation. On the other hand, a studio designed for orchestral scoring for video or film will be much larger, often with high ceilings and a greater square footage to accommodate the larger number of studio musicians.

It is not surprising that varied types of acoustical environments are found from one recording facility to the next, but it may be surprising that it is not uncommon to encounter a single facility that is capable of offering more than one environment. Certain modern studios have a multiple-room design, being made up of a central floor area with one or more *isolation (iso) booths* (Fig. 5-21) of various sizes and acoustical "aliveness," each suiting the needs of a specific set of recording applications. The iso booths are often separated from the larger central area by one or more glass sliding doors or moveable partitions. This allows the booths to be closed off from or opened up into the main area, while visual contact is maintained at all times.

Fig. 5-21. Typical isolation (iso) booth.

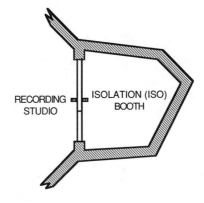

The Control Room

The *control room* of a recording facility (Fig. 5-22) serves two main purposes: to act as a critical listening environment for monitoring recorded sound over loudspeakers, and to house the equipment used in audio production.

The control room of a modern recording facility contains the audio production console, the multitrack and mastering ATRs, the production VTRs and/or VCRs, synchronizer/controller devices, effects devices, signal processors, and a specially designed amplifier/loudspeaker interface. The equipment and environment are configured so as to provide for the highest possible degree of ease of use and flexibility.

Signal-Processing Equipment in Audio-for-Video Postproduction

Over the course of the past decade, the equipment on the market that is aimed at audio-for-video postproduction uses has become increasingly flexible and sophisticated, allowing for the highly control-led creation of quality stereophonic music and soundtracks. The technological evolution of this new and thriving industry has made it vital that all audio-for-video personnel be knowledgeable about the state-of-the-art equipment available to the audio production, postproduction, and recording industries.

The following sections discuss in some detail the signal-process-ing and signal-routing equipment found in the audio-for-video pro-duction and/or postproduction facility.

Fig. 5-22. Audio-for-video control room; (A) Audio production con-sole, (B) Effects and control devices, (C) Producer's desk, (D) ARTs, (E) Monitor loud-speakers, and (F) VTRs.

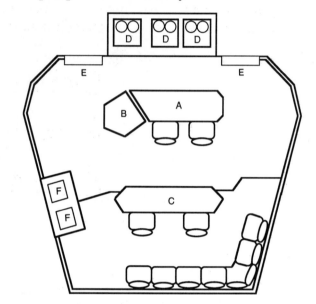

The Frequency Equalizer

Within any audio production facility, one of the most basic and impor-tant signal-processing devices is the *frequency equalizer*. This device enables the sound recordist, sound mixer, or recording engineer to ex-ercise control over the harmonic or tonal content of a recorded sound. This control may be exercised for any of a number of reasons:

- To overcome deficiencies in a microphone's frequency response
- To overcome deficiencies in a recorded track containing dialogue presence, or a musical instrument, either in the studio or in the field
- To allow contrasting sounds from several microphones or tape tracks to be better blended within a mix

- To make a recorded track sound completely different from the way it normally would, in order to achieve a particular effect
- To increase the separation between audio tracks by reducing those frequencies subject to leakage

Equalization (also commonly referred to as EQ) refers to the alteration of the frequency response of an amplifier, so that the levels of certain frequencies are more or less pronounced than those of others. Basic equalizer circuits may be broken down according to three operating modes of equalization into shelving equalizers, peaking equalizers, and high- and low-pass filters.

The simplest equalizers used by professionals are the bass and treble tone controls, which are classified as *shelving equalizers* (Fig. 5-23). Although their lack of flexibility limits their usefulness, they have some unappreciated advantages over more elaborate designs.

Fig. 5-23. Shelving equalization curves.

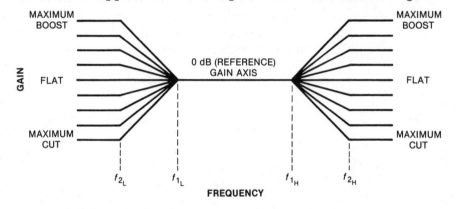

Shelving refers to a constant rise or drop in response level that continues until a selected frequency (the boost or cut level) has been reached, at which point the rise or drop tapers off. This level extends to either the high or low end of the audio spectrum on the curve and visually resembles a shelf.

The equalization curve that is most commonly found in professional audio production is that of the *peaking equalizer*. As its name implies, this form of equalizer produces a peak-shaped bell curve, which reflects either a boost or a cut at a selected center frequency (Fig. 5-24). The pole Q, or simply the Q, of a peaking equalizer refers to the width of the bell-shaped curve, which emanates symmetrically from the selected center frequency. A curve with a high Q exhibits a narrow bandwidth, containing few frequencies outside the selected bandwidth. A curve with a low Q will exhibit a broadband effect over many frequencies within the curve.

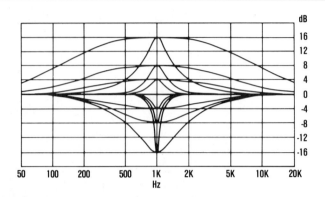

The final classification of equalizers contains the *high-pass and
low-pass filters* (Fig. 5-25). As their names imply, these filters allow
certain frequencies, which are said to be within the *passband*, to pass
through their circuits at full level, while frequencies outside the passband are attenuated. Frequencies that are attenuated by less than 3 dB
are said to be inside the passband, and those attenuated by more than 3
dB are in the *stopband*. The frequency at which the signal is attenuated
by exactly 3 dB is called the *turnover, or cutoff, frequency.*

**Fig. 5-25. High-
pass and low-
pass equaliza-
tion curves.**

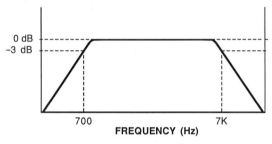

This form of equalizer is of the shelving type, with a *cut slope* of 12
dB per octave or greater. It is generally used as a subsonic filter (that is,
to eliminate warped record and floor-to-microphone rumble) or as a
high-end filter (that is, to eliminate tape, room, or record scratch
noise).

Second-Order Equalizers

Currently, there are four types of second-order equalizer systems that
are used in professional audio production. These are the selectable frequency equalizer, the graphic equalizer, the parametric equalizer, and
the notch filter.

A second-order equalization curve may be described by three interactive parameters:

1. The maximum amount of boost or cut, in dB.

2. The frequency at which the maximum amount of boost or cut occurs, in Hz.

3. The bandwidth, or Q, which is used to define the shape of the equalization curve.

The *selectable frequency equalizer* (also known as the three- or four-knob equalizer), as its name implies, has a predetermined number of frequencies from which an operator can choose. The selected frequencies provide for a continuously variable boost or cut at a predetermined bandwidth. This type of equalizer (often providing frequency selection within the low, mid-low, mid-high, and high ranges) may be found on older consoles as well as on newer, low-cost production models.

The *graphic equalizer* (Fig. 5-26) provides a series of peaking equalization curves whose center frequencies are equally spaced according to musical intervals. Therefore, an "octave band" graphic equalizer may have, for example, eleven equalization controls spaced at octave intervals, or at 20, 40, 80, 160, 320, 640 Hz, and at 1.25, 2.5, 5, 10, and 20 kHz. The equalization controls for the various bands are usually vertical linear slide controls, arranged in a side-by-side configuration. The positions of the control levers give a relative physical approximation of the actual frequency-response curve, and thus the name graphic equalizer. This type of equalizer is often used in applications requiring the fine tuning of a system to match the acoustics of a given room, such as an auditorium or a control room.

Fig. 5-26. Rane GE 30 graphic equalizer.
(Courtesy of Rane Corporation.)

The third type of second-order equalizer is the *parametric equalizer* (Fig. 5-27). The name is meant to imply that the center frequency of each band is continuously adjustable, rather than switchable by steps. Control over the center frequency bandwidth, or Q, is often also continuously variable, although some manufacturers provide for switching between two discrete values. Finally, the amount of boost or cut is usually continuously variable with this type of equalizer. Generally, each band of continuously variable frequencies will overlap and cascade into the next band, in order to provide for smooth transitions between bands or to allow for multiple high-Q peak/dip equalization within a narrow frequency range.

Fig. 5-27. Orban 622B parametric equalizer.
(Courtesy of Orban Associates, Inc.)

During the last decade, parametric equalizers have become the standard in the input strips of most modern recording and audio production consoles. This popularity is due to the flexibility and performance capabilities they exhibit, while maintaining a small size.

Another type of second-order equalizer, known as a *notch filter*, may be used for the modification of sound, as well as for the removal of hum and other undesirable noises of discrete frequency. This filter may be tuned to a very narrow bandwidth, or Q, to attenuate a particular frequency (Fig. 5-28). Problem frequencies (often the hum of the power main at 50-60 Hz) are eliminated, with little or no effect on the overall program sound. Notch filters are often used within the video edit and sweetening bays, because of the need for filtering work on original film and EFP and ENG field audio tracks.

Fig. 5-28. High-Q response of a notch filter.

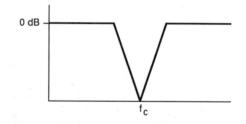

Automatic Gain Controllers

The dynamic range of music is on the order of 120 dB, while that encountered in the recording and broadcast media is generally considerably less than this figure. Whenever an audio program that is recorded, reproduced, or distributed to the public has an excessively wide dynamic range, two things may possibly happen to the audio signal: quiet passages will be drowned out by tape or system noise or by the ambient noise of a listening area (35−45 dB is the sound pressure level for the average home); and loud passages will either be distorted within the system's audio chain or will be reproduced at too loud a level. Similarly, if a program of wide dynamic range is reproduced through a medium with a narrow dynamic range, such as an AM radio (20–30 dB) or an FM radio or stereo television (40–50 dB), a great deal of audio information is lost in the background noise.

To prevent these problems, the dynamic range of audio program material may be reduced to a level that is appropriate both for the medium through which it is to be reproduced and for comfortable listening in the average home. This reduction of the overall audio dynamic range may be accomplished through the manual riding of faders by the recording engineer or through the use of an automatic gain controller.

Professional automatic gain controllers, also known as dynamic range changers, can be broken down into four categories: the compressor, the limiter, the expander, and the noise gate.

The Compressor

The *compressor* is, in effect, a variable amplifier that operates as an automatic fader. When an incoming signal exceeds a predetermined level, known as the *threshold* of the device (Fig. 5-29), the gain is automatically reduced, and the signal output is attenuated in proportion to the intensity of the input signal. The increase in input signal level (rated in decibels) that is required to cause an increase of 1 dB at the output signal is called the *compression ratio*, or the slope of the compression curve. Given a compression ratio of 8:1, a post-threshold increase of 8 dB at the input will produce an increase of 1 dB at the output. Fig. 5-30 shows an example of a compressor, with its associated compression curve.

Fig. 5-29. Typical compression curves (input level vs. output level) of the UREI 1176 LN. *(Courtesy of JBL, Inc.)*

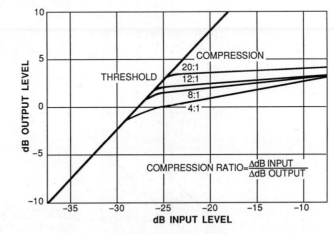

Since the level of sound generated by musical instruments and voices will vary in loudness, input signals may register above the threshold of the device at one instant and below it the next. The length of time it takes for the gain to be restored to unity once the input signal falls below the threshold is respectively determined by the *attack time* and *release time* of the compressor.

Fig. 5-30. EMT 261 compact compressor/ limiter. *(Courtesy of Gotham Audio Corporation.)*

(A) Controls.

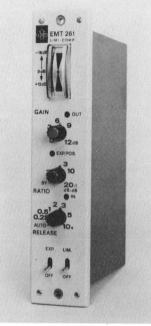

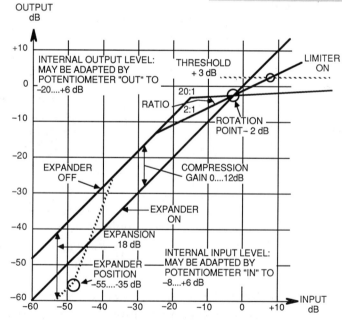

(B) Static characteristics.

The perception by the ear of the loudness of a signal is proportional to its rms level (room mean square, or average, value). Thus, high-level peaks of short duration do not noticeably increase the overall loudness of a signal. If these peaks of a waveform were permitted to trigger gain reduction, the output volume would actually decrease rather than increase, perceptibly changing the dynamics of the program material. To avoid the triggering of compression by short-duration peaks, the attack time is set such that the waveform must exceed the threshold for a long enough time to effect an increase in the rms level of the sound. The attack time is therefore defined as the time it takes for the gain to decrease to a specific amount, usually to 63% of its final value.

If the release time of the compressor is set at a period that is too short to suit the program material (that is, full gain will be restored each time the signal falls below the threshold), audible thumps, "pumping," and breathing may be audible as a result of the rapid rise of background noise as the gain is increased. Also, if a rapid succession of peaks is fed into the device, the gain will be restored after each one, and the sound level of the program will be heard rising after each peak. To eliminate these effects, longer release times may be used to allow the compressor to retain its level of gain reduction throughout short-term, or transient, deviations of the signal and to return to normal operation gradually, once the input signal has consistently returned to levels below threshold. A longer release time may serve to reduce pumping. However, if the release time is too long, a loud section of the program may cause a gain reduction that will carry over into a softer musical or vocal passage, making the softer passage less audible or even inaudible. The release time is defined as the time needed for the output signal of the device to increase to 63% of its full gain.

The compressor may also incorporate other functions for controlling dynamic range. Including equalization within the detection circuitry of a compressor yields a frequency-selective compression device known as a *de-esser* (Fig. 5-31), which may be used to reduce the high-end sibilance (distorted sound of the letter *s*) associated with a recorded vocal track.

Many of the more recent compressors are also equipped with a feature for the *ducking* of background music (or other program material). With this option, one signal source (such as the music track) is passed through the device's normal audio path, and a second signal is connected to a side-chain input. The controls may be then set so that the presence of the signal at the side-chain input (the voice of the narrator) will cause a specific gain reduction between 0 dB and approximately 50 dB to occur on the normal audio path (the background music). Thus, the music is "ducked" whenever the narrator is speaking. Within the audio-for-video production facility, such interactive gain control may be invaluable in achieving a tight dynamic balance between vocal and musical tracks.

Fig. 5-31. dbx 902 de-esser.
(Courtesy of dbx, a Division of BSR North America, Ltd.)

The Limiter

When the gain reduction for a compressor is a large enough value, the device is known as an output limiting device, or a *limiter* (Fig. 5-32). A limiter is used to prevent average and/or peak signals from exceeding a predetermined threshold level. This device is used in the recording, video, and broadcast media to prevent the overloading of audio program material as output from production consoles, magnetic tape players, amplifiers, and broadcast transmitters.

Most limiting devices operate at output ratios of 10:1 or 20:1, but many are available with ratios approaching 100:1. Since such a large increase in level is necessary at the input of the limiter to produce even a small increase in its output, the use of such devices greatly reduces the likelihood of equipment overload.

Limiting is most often required during the recording of program material to prevent short-duration peaks from reaching their full amplitude. In practice, these peaks add little information to the program material relative to the distortion they would cause if allowed to satu-

Fig. 5-32. EMT 257 compact limiter *(Courtesy of Gotham Audio Corporation.)*

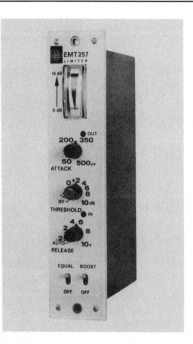

(A) Controls.

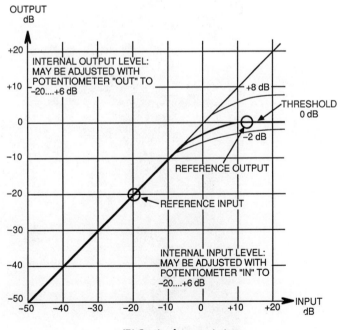

(B) Static characteristics.

rate the tape or the noise that would be induced if the overall signal were recorded at a lower level.

Short attack and release times may act to reduce high-level transients (sharp-attack waveforms) in such a way that the ear cannot detect the gain reduction. The limiting process should be employed only to remove occasional peaks, since gain reduction due to numbers of peaks in quick succession would be too noticeable. Therefore, if the program material contains many peaks, the threshold should be raised and the overall gain reduced manually so as to limit only the occasional extreme ones.

The Expander

Expansion is the process of increasing the dynamic range of audio material. This may be accomplished through the use of an *expander* (Fig. 5-33), which is capable of automatically decreasing the gain as the signal level falls below a threshold or increasing the gain as the signal level rises above the specific threshold. Thus, expanders increase the overall dynamic range of program material by making loud signals louder and/or soft signals softer.

Fig. 5-33. Kepex II expander.
(Courtesy of Valley People, Inc.)

Within audio recording and postproduction, the expander has found many applications, for example, in the reduction of leakage from one microphone to another in close-miked configurations or in the reduction of background noise (tape or ambient) behind the desired recorded track. The expander, in many cases, acts as a noise reduction device by reducing the level of unwanted signals (those that lie below a carefully selected threshold) while preserving the desired signal (which falls above the threshold point).

The Noise Gate as Expander

Another type of expansion device is the *noise gate* (Fig. 5-34). This device acts as an infinite expander, allowing input signals that lie above any selected threshold to be passed through to the output at unity gain and without dynamic processing. However, if an input signal falls below the threshold level, the noise gate effectively shuts out the signal by applying full attenuation to the output. In this way, desired signals are allowed to pass while unwanted noise is not. Noise gates are very effective for the reduction of leakage when drums or other instruments are miked with multiple pickups.

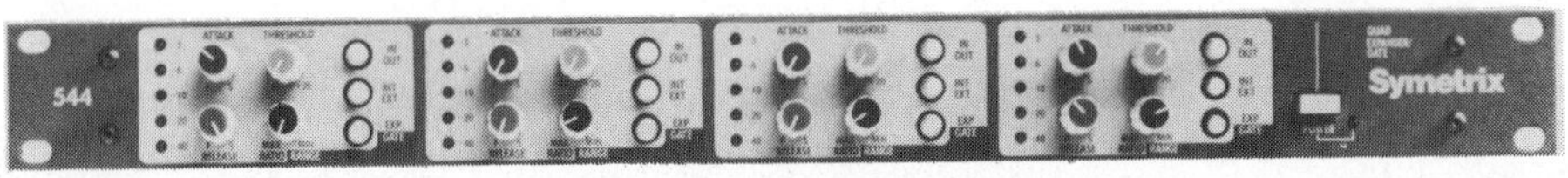

Fig. 5-34. Symetrix 544 noise gate. *(Courtesy of Symetrix.)*

Some noise gates are designed so that a special *key input* (Fig. 5-35) can be fed into the device. This allows a chosen analog signal source, such as a miked instrument, a synthesizer or an electronic oscillator, to trigger the output of the gate. In this way, a separate signal can be used as a controller of an audio signal whose level is dictated by the output of the noise gate.

Fig. 5-35. Diagram of a basic keyed-input noise gate.

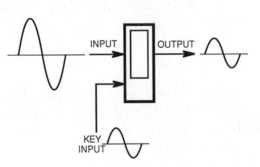

Modular Signal-Processing Systems

In recent years, several manufacturers of signal-processing equipment have adopted modular designs. Compact signal-processing modules can be fitted into a portable or standard 19-inch rack-mounting frame. The idea behind the modular approach is that users are able to tailor a system to meet specific current requirements, and it will still have enough flexibility and expandability to meet their future needs. One modular system is SCAMP (Standardized Compatible Audio Modular Package) from Audio + Design/Calrec, Inc. (Fig. 5-36). The SCAMP system was designed to house a wide range of signal-processing equipment in a space-saving, high-density format. As many as seventeen 1-inch modules may be accommodated in a single 19-inch frame, with several modules having a dual-channel capability. The following modules are available for the SCAMP system: expander/gate, dual noise gate, dynamic filter noise gate, de-esser, compressor/limiter, sweep equalizer, parametric equalizer, octave equalizer, crossover/4-band processor, mic preamp, distribution amp, delay/time shape, and panning effects.

Fig. 5-36. The SCAMP (Standardized Compatible Audio Modular Package) system. *(Courtesy of Audio + Design/ Calrec, Inc.)*

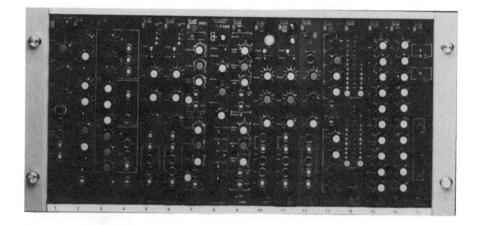

The dbx 900 Series (Fig. 5-37) is another modular signal-processing system. It offers the user the option of including up to eight signal-processing modules within a rack-mountable frame that is 5-1/4 inches high. At present, the 900 Series offers three modules: the Model 902 de-esser, the Model 903 compressor, and the Model 904 noise gate.

**Fig. 5-37. The
dbx 900 Series
modular signal-
processing sys-
tem.** *(Courtesy of
dbx, a Division of
BSR North Amer-
ica, Ltd.)*

Noise Reduction Devices

With the increasing demand for quality audio for video, it is necessary
to pay close attention to the background noise levels produced by magnetic tape. The overall dynamic range perceived by the human ear en-
compasses 130 dB, but this range from the softest perceptible sounds
to the loudest cannot be adequately recorded on magnetic tape without
imposing some degree of dynamic range compensation (Fig. 5-38).
This limitation of conventionally recorded ATR and VTR analog audio
tracks is due to tape noise, which is perceptible when the overall sig-

**Fig. 5-38. Dy-
namic ranges of
several audio
devices.**

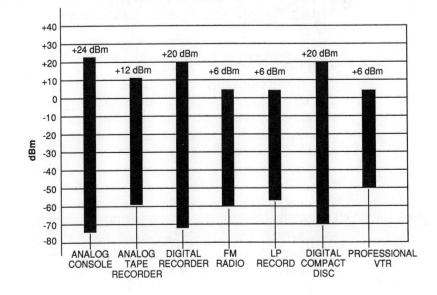

nal level is too low or as a result of distortion caused by tape saturation at excessively high recording levels.

If the optimum recording level produces an unacceptable amount of noise, the engineer or operator is faced with a choice between two options: to record at a higher level (with the possibility of yielding an increased distortion) or to change the overall dynamic range of the signal.

The types of noise that must be eliminated from an audio track may be classified as follows: tape noise, crosstalk between tracks, print-through, and modulation noise. These analog-based noises may be reduced or eliminated by several means. Recording tape may be chosen for its reduced noise level and/or the tape speed may be increased in order to record at higher flux levels. The tape heads and electronics may be of such design that very little crosstalk exists between recording channels. Tape with a thicker base layer may be used to achieve a reduction in print-through. In fact, the availability of low-noise, high-output tapes (which have improved on the average signal-to-noise ratio of professional audio tapes by 3 dB) and better ATR and VTR designs have greatly reduced system noise levels in general.

Although tape noise may no longer be such a limiting factor in standard day-to-day audio production, the addition of added tracks in multitrack production brings with it an associated increase in tape noise. During the past decade, several means of reducing the noise inherent in the analog recording medium have been developed: the compander noise reduction system, the single-ended noise reduction system, the noise gate, and modifications in tape formulation and speed.

The Compander System

The most commonly encountered noise reduction device used within analog audio recording chains is the *compander system*. These systems, such as the Dolby® and dbx® ones, are currently utilized in professional and nonprofessional audio production applications worldwide.

The compander system gets its name from the fact that during the process of encoding the recorded signal onto tape, the device functions as a compressor, and during the process of decoding the noise-reduced signal off the tape, the device exhibits a reciprocal expansion characteristic (Fig. 5-39). Through this method of noise reduction, the overall dynamic range of the signal recorded onto tape is compressed into a restricted and tightly controlled range. As a result, passages made up of low-level signals can be recorded on tape at a higher operating level. On playback of the recorded sound, the compressed signals are expanded back up to their original dynamic range in a precisely controlled reciprocal process. During this process of expanding low-level passages back to their original dynamic level, the noise associated with analog tape is reduced from the original level to a less perceptible one.

Fig. 5-39. The compansion noise reduction process.

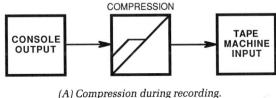

(A) Compression during recording.

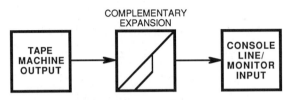

(B) Reciprocal expansion during playback.

In addition to the reduction of the noise associated with audio tape, companders may also be useful in reducing extraneous noise and hum originating in long-distance transmission or broadcast lines.

The Dolby® Noise Reduction System—The Dolby noise reduction system is available as two types of companders: Dolby® A and Dolby® B.

The Dolby A system is the professional model, providing 10 dB of noise reduction below 5 kHz and a maximum of 15 dB at 15 kHz. In this system, compression is achieved by increasing the gain of low-level signals rather than decreasing the gain of high-level signals. The circuitry includes a limiter that prevents the signal from rising much above −40 dBm at its output. The output of this limiter is added algebraically to the uncompressed input signal. Since the output of this limiter will only ever rise slightly above −40 dBm, the effect of the addition of this output to the input signal depends on the level of the input signal. When the input signal is low, the output of the limiter is large in comparison and its addition to the input signal results in an overall level boost. When the input signal is high, the contribution of the limiter's output is negligible when the two signals are added.

Expansion is accomplished in exactly the same way except that the output of the limiter is subtracted from the input signal, thus reciprocally reducing the gain at lower levels.

In order to effectively reduce the entire noise spectrum, program audio is divided into four separate bands within the Dolby A system. Each band of frequencies has its own dynamic range processor, so the presence of a high-level signal in one band does not interfere with detection in the other bands. The four frequency bands are: (1) 80 Hz low-pass (80 Hz and below), (2) 80 Hz to 3 kHz bandpass, (3) 3 kHz high-pass (3 kHz and above), and (4) 9 kHz high-pass.

The Dolby B system, commonly found in consumer cassette decks, is specifically designed to reduce tape hiss. This system compands only the upper-frequency components and has no effect on lower-frequency noise such as hum or rumble. The Dolby B system provides noise reduction of 3 dB at 600 Hz; this figure rises to 10 dB at 5 kHz, at which point it levels off in a shelving fashion.

The dbx® Noise Reduction System—The dbx 180 Type I noise reduction system (Fig. 5-40) is a compressor/expander system that is interfaced into a recording system in precisely the same way the Dolby system is. The dbx compander system is able to provide for up to 30 dB of noise reduction, yielding an overall analog signal-to-noise ratio on the order of 90 dB for a professional ATR or VTR. Dynamic range processing in this device operates at a 2:1 ratio between −90 dBm and +25 dBm, with unity gain occurring at 0 VU.

Fig. 5-40. The dbx 180 Type I noise reduction system. *(Courtesy of dbx, a Division of BSR North America, Ltd.)*

Unlike the Dolby noise reduction system, the dbx system compands over the entire frequency range, from 20 Hz to 20 kHz, resulting in a lower degree of interdependence between device and signal level. Low-frequency rumble, which could act to offset and distort full-band compander operation, is avoided to a certain extent through the use of a filter that preemphasizes high frequencies by 12 dB before the input signal enters the compressor. On decoding, the signal is passed through a reciprocal cut filter to deemphasize the highs.

The Single-Ended Noise Reduction System
The Dolby and dbx noise reduction systems are also known as double-

ended noise reduction devices because they serve to reduce the noise engendered within a storage or transmission medium. Noises existing outside this medium cannot be prevented from entering, only distortions inherent within the medium itself are affected. Only through the use of single-ended, or noncomplementary, noise reduction devices can signals, such as those of a recorded audio track, be dynamically noise processed.

Basically, a *single-ended noise reduction system* (Fig. 5-41) operates as a frequency-dependent expander or noise gate, deleting unwanted noise from audio source material by utilizing a downward dynamic range expander in conjunction with a program-controlled, dynamic low-pass filter. The expander and filter may operate in tandem or separately in order to provide the greatest possible amount of noise reduction.

Fig. 5-41. Symetrix 511, a single-ended noise reduction system. *(Courtesy of Symetrix.)*

The expander, which consists of a voltage-controlled amplifier and associated detection circuitry, selectively reduces noise of a specific frequency by attenuating the signal whenever the level falls below an adjustable threshold point. This results in an increased dynamic range over the entire audio spectrum. The filter accomplishes further noise reduction by taking advantage of two basic psycho-acoustical principles:

1. Sound is capable of masking noise within the same bandwidth.
2. Reducing the bandwidth of an audio signal reduces the perception of noise (the greater the spectral distribution of noise, the greater is the sensitivity of the human ear to that noise).

The dynamic filter examines the incoming signal for high-frequency content. In the absence of high frequencies, the signal bandwidth is reduced to as little as 2 kHz. Once the high-frequency content returns, the passband returns to unity gain, passing the entire signal.

The Noise Gate
Like dynamic noise reduction devices, noise gates may often be very effective in the reduction of background noise. When used for this pur-

pose, a noise gate may be viewed as a "high ratio" expander. That is, the device acts as a unity gain amplifier, passing all signals whose level is above a specific threshold. When the input signals falls below the predetermined threshold level, the output signal is effectively shut down. Thus, the noise gate can function as an infinite noise reduction device.

Certain program material may call for additional fine tuning of the noise gate's attack and release controls in order to avoid excessive breathing or pumping of the noise floor below the desired signal.

Modifications in Tape Formulation and Speed

With recent improvements in the formulation of professional audio recording tape, many within the audio community have abandoned the use of multitrack noise reduction devices in favor of low-noise, high-output tapes. The availability of these new formulations reduced the analog machine's average signal-to-noise ratio by 3 dB. Most users of these tapes have also opted to increase recording speed to 30 ips, which produces a further drop in the signal-to-noise ratio and an improvement in transient response.

Audio Effects Devices

In addition to the devices for equalization, automatic gain control, and noise reduction that are available to the audio-for-video sound recordist, sound mixer, and engineer, a complete range of devices exists that may be used to change, emphasize, or enhance an already existing signal. Such devices have come to be termed *audio signal processors* or *audio effects devices*. Although these devices were originally designed for and continue to be utilized in music production, the adaptation of the music studio for audio-for-video production has brought about the adoption of these program-enhancing devices as a matter of course. The effects devices that are encountered in the field of audio-for-video production include the reverberation unit, the digital delay line, the pitch changer or harmonizer, the time compressor/expander, microprocessor-based effects devices, and digital sampling devices.

The Reverberation Unit

In professional audio production, the most important tool for the enhancement of sound is the reverberation unit. Reverberation, by definition, is the persistence of a sound due to the perception of closely spaced and random echoes reflected from one boundary to another within an enclosed space. This natural effect gives perceptible and imperceptible cues as to the size, density, and nature of that space and adds to the perceived warmth and depth of recorded sound.

The reverberated sound itself may be broken down into three sub-components: the direct signal, the early reflection, and the reverberation time (Fig. 5-42). The *direct signal* is the original sound wave received by the listener. The *early reflection* is made up of the first few reflections of that sound that are projected to the listener from major boundaries within a determined space. It is generally these reflections from which the listener subconsciously receives cues as to the size of a space. The last signals to be perceived make up the characteristic of *reverberation time*. These signals may be broken down into many random reflections going from boundary to boundary within the space. These reflections are so closely spaced in time that the brain is unable to discern any individual reflection and will perceive them as one steady and decaying signal. The reverberation time (often referred to as T 60) may be defined as the time required for all signals following the direct signal to decay to a level 60 dB below the initial intensity.

Fig. 5-42. Signal level vs. time for reverberation.

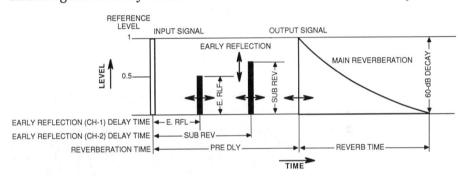

Currently, a reverberation device that is gaining wide acceptance throughout the professional audio industry is the *digital reverberation unit*. The digital reverb unit is a microprocessor-based effects device that has a great degree of control over parameters affecting the reverb quality, such as level, decay time, pre-echo delay, low EQ, high EQ, and variable EQ crossover points. This device gives the operator the ability to preset a wide variety of parameters by programming values into computer memory, allowing for specific reverb-type effects to be instantly changed. Many of these parameters may be predefined so that the operator can call up, at will, such familiar reverb "programs" as a large concert hall, a small hall, or plate and spring reverb simulations.

Digital reverberation is accomplished through the regeneration of an input signal as a series of closely spaced digital delays. Through the application of complex digital algorithms upon the signal, the delays are programmed to follow a well-defined set of random patterns, resulting in a dense reverberation of a specific and selected character.

A digital reverberation unit that is widely used throughout the recording and broadcast industries is the Lexicon 224X digital reverberator (Fig. 5-43). Using complex algorithms, this device simulates the acoustics of various architectural spaces and produces a variety of special effects, including simulated reverberation programs for rooms and echo chambers, as well as plate, spring, and inverse room simulations. Effects programs for the 224X include chorus, echo, resonant chords, and multiple-band delays. Users are able to choose from 18 factory-preset programs with 59 reverb and effects variations, or they may create and store their own programs using the unit's keyboard. Each of these programs may be split over four separate tracks, creating two independent, full-bandwidth, stereo reverb and effects programs. The associated Lexicon Alphanumeric Remote Control (LARC) unit allows for the storage of 36 preset programs, which may be offloaded onto cassette for storage and/or backup. Frequently used programs may be quickly addressed by means of eight independent keys on the LARC. A numeric keypad addresses all programs, variations, and registers, and the main display gives a readout of all associated program and parameter information in alphanumeric form.

Fig. 5-43. Lexicon 244X digital reberverator with LARC.
(Courtesy of Lexicon, Inc.)

The Digital Delay Line
One of the newer but commonly encountered effects devices for musical and vocal special effects is the *digital delay line* (Fig. 5-44), commonly referred to by the abbreviation *DDL*. Indeed, the use of this device has practically become the signature of modern music produc-

tion. In its standard operating mode, the digital delay line serves to introduce a series of one or more discrete repetitions of the input signal at certain regularly spaced, user-defined intervals of time (Fig. 5-45).

Fig. 5-44. Lexicon PCM 41 and PCM 42 digital delay processors. *(Courtesy of Lexicon, Inc.)*

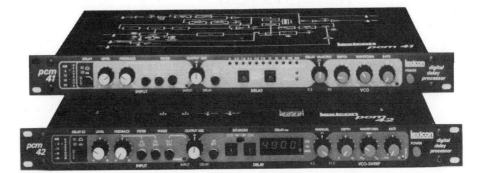

Fig. 5-45. The digital delay line introduces one or more discrete repeats of the input signal at user-defined intervals.

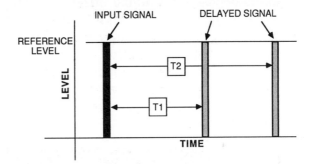

A practice that has become common since the advent of the DDL is *automatic double tracking*. The recorded signal to be processed is electronically "doubled" by the introduction of one or more delays. In this way, a fuller and more expansive sound may be achieved. A musical ensemble can be made to sound as though the number of participants were greater. Other special effects attained through the use of the digital delay line are slap echo, electronic reverb, infinite repeat, time sweeps, and flange effects. Certain DDLs, such as the Lexicon PCM 42, incorporate a metronomic feature, allowing the user to synchronize music and delay with precise rhythmic control.

In operation, the digital delay line encodes the analog input signal into a PCM digital form. This binary signal is then fed into a shift register (a device for the temporary storage of digital information), where it can be stored for a specific duration of time that is determined by the memory capacity of the device. Next, a high-frequency clocking oscillator reads the signal information out of shift register at a user-defined delay interval. This signal is then reconverted back into analog form, through the use of a D/A converter, and delivered to the output of the DDL.

The Pitch Changer

Depending on the application for which a digitally based *pitch changer* was designed, this type of effects device has one of two basic operating functions: (1) to change the pitch of an audio source without affecting the program duration, or (2) to change the duration of audio program material without affecting the original pitch. It is obvious that pitch changers are very useful to production teams in both music score recording and audio-for-video postproduction.

For transposing the relative pitch of an audio source without affecting the program duration, the most widely used pitch changer is the Eventide H969 ProPitch Harmonizer® (Fig. 5-46). This device acts simultaneously as a digital delay line and pitch changer, with a delay range of up to 1530 milliseconds at full bandwidth and a maximum pitch range of 1 octave up and 3 octaves down. The Harmonizer® allows for continuous control over either pitch or delay through the use of dual-concentric pots for coarse and fine adjustment. The pitch ratio may also be varied up or down by the precise increment of a major third, minor third, fifth, seventh, or octave through the use of instant pitch change presets.

Fig. 5-46. Eventide H969 Pro-Pitch Harmonizer®. *(Courtesy of Eventide, Inc.)*

The Time Compressor/Expander

The *time compressor/expander* is currently used in video postproduction, film, and broadcasting to make changes in the running time of a film or video and audio recording while maintaining the original, natural pitch of voices, music, and effects. This device may be used to either shorten or lengthen a television or radio spot so it will fit into an allotted time slot.

The Lexicon 2400 stereo audio time compressor/expander (Fig. 5-47) is designed to interface with time code synchronizers or directly with tape machines that have the capability for time code following, thereby directing time compression or expansion in sync. This machine is designed to operate in two possible modes: as master and as slave. In the master mode, the machine can speed up or slow down the tape transport in accordance with original and modified running times entered via the control panel. In the slave mode, the machine measures the speed of the transport, corrects the pitch shift, and reports the speed factor on the front panel's alphanumeric display. Basic specifications for the 2400 include ten storage registers (or memory groups) featuring nonvolatile memory, a speed factor ranging from 0.750 to 1.333 times the original recording speed, long-term accuracy of 2 frames over a 1-hour period, and an overall lockup time typically less than 5 seconds.

Fig. 5-47. Lexicon 2400 stereo audio time compressor/expander.
(Courtesy of Lexicon, Inc.)

Microprocessor-Based Effects Devices

In recent years, the professional audio market has witnessed the introduction of *microprocessor-based effects devices*. These extremely versatile systems, which process digitally encoded PCM audio signals using computer-generated algorithms, are able to change parameters quickly in order to operate as multiple-function effects devices.

One such device is the Eventide SP 2016 signal processor (Fig. 5-48). This unit is equipped to operate up to 19 user-programmable processors (contained in individual LSI chips) and has the capacity to store up to 65 user-created programs. As many as eleven ROM integrated-circuit software programs may be interchangeably inserted into the unit at once. Available software programs include six individual reverb programs, LosslessRoom® (allowing layer upon layer of sound without decay), automatic panner, channel vocoder (voice-operated coder), loop editing, multitap delay, time scramble, chorus, flanging, four frequency band delay, musical comb filter, digiplex (simulates tape head echo), and delay. A unique capability of the SP 2016 is that it may be interfaced with an IBM PC/XT (or compatible machine) or a

Hewlett-Packard personal computer, allowing the user to design or modify his or her own software. This integrated software development system is intended for operators who have general knowledge of and experience with audio signal processing. No specific knowledge of advanced programming techniques is required.

Fig. 5-48. Eventide SP 2016 signal processor.
(Courtesy of Eventide, Inc.)

The Lexicon PCM 70 digital effects processor (Fig. 5-49) innovatively incorporates the Dynamic MIDI (Musical Instrument Digital Interface) into its operating structure. Designed for both electronic music scoring and the recording studio, the PCM 70 applies the concept of the plug-in LSI chip, allowing for additional software to be added to the system as new programs become available. Presently available programs include concert hall, rich chamber, rich plate, infinite reverb, resonant chord, multiband delay, chorus, and echo effects. With the Dynamic MIDI, as many as 10 parameters for effects or reverb may be assigned to any MIDI-connected controller or keyboard, with partial control being assignable to key velocity, pressure, or after-touch. More than 80 different types of parameters, such as delay time, beats-per-minute, feedback, wet-dry mix, high-pass and low-pass filters, room size, etc., are specified within the available programs. Patches may be assigned internally or to external devices via MIDI protocol. In addition, this system incorporates a register table, which allows all of the 128 MIDI-specified presets to be utilized in recalling program and user registers.

Fig. 5-49. Lexicon PCM 70 digital effects processor.
(Courtesy of Lexicon, Inc.)

Digital Sampling Devices

Digital sampling refers to a device's ability to digitize and store audio signals within a digital memory and to recall this sound from memory upon an external trigger signal. Digital sampling devices may be sub-divided into three categories: drum machines, triggered DDLs or DSPs, and pitch-controllable samplers.

The *drum machine* was one of the first sampling devices to find widespread application. Recorded digitally within the permanent and user-accessible memory of these devices are the individual sounds of the modern drum kit. These sounds may be pulled from the memory through the use of real-time triggers, such as push buttons or drum pads, or the sounds may be programmed to play in stepped time via an electronic sequencer. Most modern drum machines may be MIDI-inter-faced for multiple-instrument control using a sequencer.

Triggered DDLs or DSPs are digital delay lines or digital signal pro-cessors that have the ability to store a sampled signal within memory and to recall this signal upon an external trigger. In general, only one sound or sound mixture may be sampled and reproduced at a time; however, newer models of these devices, such as Advanced Music Sys-tems DMX 1580-S and the Publison Infernal Machine 90, have the use-ful feature of a sample-editing facility for tailoring the beginning and end of the sampled signal. The Publison Infernal Machine 90 (Fig. 5-50), in addition to being a versatile microprocessor-based effects de-vice, is capable of random access multiple sampling at full bandwidth under MIDI protocol.

Fig. 5-50. The In-fernal Machine 90, a stereo audio computer. *(Courtesy of Publison America, Inc.)*

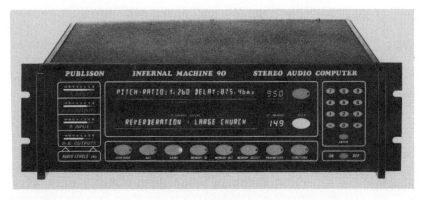

The category of *pitch-controllable samplers* includes sampling in-struments under the control of musical keyboards. These devices, such as the Fairlight CMI, the Synclavier III (Fig. 5-51), the Kurzweil 250, and others, allow for extensive manual, automated, and MIDI control over a sampled signal. Functions include user sampling, extensive editing control over the sample, note transposition, polyphonic voic-ings, tailoring of start and end times, looping, and reverse playback.

Fig. 5-51. The Synclavier Digital Music System. *(Courtesy of New England Digital Corp.)*

The Audio Production Console

At the heart of audio editing, sweetening, or recording is the *audio production console*. The function of this device is to provide for complete and flexible control over tone, blending, processing, and spatial positioning of signals fed to its inputs by microphone (mic and direct signal feeds) and line level devices (ATRs, VTRs, effects devices, electronic musical instruments, and various other input and output line feeds). The console must be able to route these processed signals to any appropriate combination of other devices, tape machines, and monitor systems, so that the recording, sweetening, and mixing functions can be performed as quickly and reliably as possible. In that the audio production console makes possible subtle combinations and mixtures of electrical information that would otherwise remain rigidly fixed and uncontrolled, it may be said to be to the engineer or sound mixer what the palette is to the painter. It is the console that allows the recording engineer to experiment, to blend and control the wide range of variations that may occur in the audio portion of a program.

It is important to point out that audio production consoles may be designed for a variety of services. Their range of applications include live ENG and EFP sound pickup, video and audio editing in postproduction, audio production for video, and music recording. Although these systems differ in appearance, location of controls, and methods of signal routing, automation, and storage (depending on application and manufacturer), the overall layout and capabilities will generally be similar. This means that the operator who has a functional knowledge of console operation and design can approach any production system with a relative degree of confidence.

An example of an audio production console is shown in Fig. 5-52.

Fig. 5-52. SSL
6000 E console.
*(Courtesy of Solid
State Logic, Ltd.)*

The I/O Module

In their construction, most modern professional audio consoles incorporate an input strip module, known as an *input/output, or I/O, module* (Figs. 5-53 and 5-54). The I/O module is an interchangeable, plug-in strip, which includes all of the necessary electronics to functionally process a recorded signal, including mic/line preamplifier, equalizers, auxiliary send controls, record monitor controls, and channel assignment matrix. The location of all electronics and immediate switching on one circuit board allows for reduced production costs, ease of serviceability, and custom configuration within a standard mainframe design.

Fig. 5-53. The I/O module of the SSL 6000 E.
(Courtesy of Solid State Logic, Ltd.)

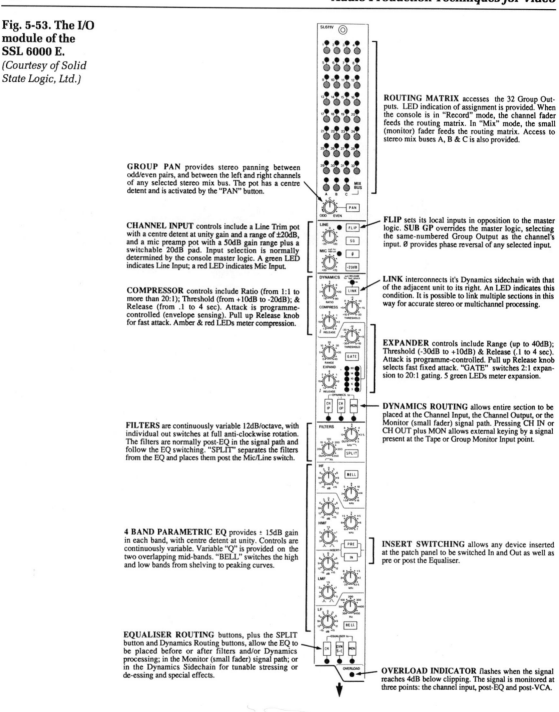

GROUP PAN provides stereo panning between odd/even pairs, and between the left and right channels of any selected stereo mix bus. The pot has a centre detent and is activated by the "PAN" button.

CHANNEL INPUT controls include a Line Trim pot with a centre detent at unity gain and a range of ±20dB, and a mic preamp pot with a 50dB gain range plus a switchable 20dB pad. Input selection is normally determined by the console master logic. A green LED indicates Line Input; a red LED indicates Mic Input.

COMPRESSOR controls include Ratio (from 1:1 to more than 20:1); Threshold (from +10dB to -20dB); & Release (from .1 to 4 sec). Attack is programme-controlled (envelope sensing). Pull up Release knob for fast attack. Amber & red LEDs meter compression.

FILTERS are continuously variable 12dB/octave, with individual out switches at full anti-clockwise rotation. The filters are normally post-EQ in the signal path and follow the EQ switching. "SPLIT" separates the filters from the EQ and places them post the Mic/Line switch.

4 BAND PARAMETRIC EQ provides ± 15dB gain in each band, with centre detent at unity. Controls are continuously variable. Variable "Q" is provided on the two overlapping mid-bands. "BELL" switches the high and low bands from shelving to peaking curves.

EQUALISER ROUTING buttons, plus the SPLIT button and Dynamics Routing buttons, allow the EQ to be placed before or after filters and/or Dynamics processing; in the Monitor (small fader) signal path; or in the Dynamics Sidechain for tunable stressing or de-essing and special effects.

ROUTING MATRIX accesses the 32 Group Outputs. LED indication of assignment is provided. When the console is in "Record" mode, the channel fader feeds the routing matrix. In "Mix" mode, the small (monitor) fader feeds the routing matrix. Access to stereo mix buses A, B & C is also provided.

FLIP sets its local inputs in opposition to the master logic. **SUB GP** overrides the master logic, selecting the same-numbered Group Output as the channel's input. Ø provides phase reversal of any selected input.

LINK interconnects it's Dynamics sidechain with that of the adjacent unit to its right. An LED indicates this condition. It is possible to link multiple sections in this way for accurate stereo or multichannel processing.

EXPANDER controls include Range (up to 40dB); Threshold (-30dB to +10dB) & Release (.1 to 4 sec). Attack is programme-controlled. Pull up Release knob selects fast fixed attack. "GATE" switches 2:1 expansion to 20:1 gating. 5 green LEDs meter expansion.

DYNAMICS ROUTING allows entire section to be placed at the Channel Input, the Channel Output, or the Monitor (small fader) signal path. Pressing CH IN or CH OUT plus MON allows external keying by a signal present at the Tape or Group Monitor Input point.

INSERT SWITCHING allows any device inserted at the patch panel to be switched In and Out as well as pre or post the Equaliser.

OVERLOAD INDICATOR flashes when the signal reaches 4dB below clipping. The signal is monitored at three points: the channel input, post-EQ and post-VCA.

Fig. 5-53 (cont.)
The I/O module
of the SSL 6000 E.
(Courtesy of Solid
State Logic, Ltd.)

STEREO CUE SEND provides panning and level control. Send source may be pre or post either the large (channel) fader or the small (monitor) fader. Level pot incorporates a push-push switch to turn send on & off.

CUE/AUXILIARY SENDS have individual level controls with push-push on/off switches. Each send source may be selected pre or post large or small fader.

GROUP TRIM is the submaster for same-numbered output groups (bus). Range is -30dB to 0dB; a detent is provided at unity. DIRECT routes the channel output directly to it's same-numbered bus.

FLOAT removes the channel output from any Mix Bus to which it is assigned, and sends it to the routing matrix for further assignment. Uses include patchfree audio subgrouping and track bouncing.

MULTI-TRACK REMOTES are built-in to allow one button drop-ins. Either READY GROUP or READY TAPE switches the associated track from Safe to Ready. simultaneously selecting Tape or Group as the source. When both are pressed, Tape+Group is monitored until Record is entered; the monitor source then switches automatically to Group.

SMALL FADER serves as multitrack monitor when the console is in Record mode, feeding the selected mix bus via the L/R pan pot. Solo is normally In-Place, but may be switched to AFL (non-destructive) by master.

In Mix mode, small fader feeds routing matrix. Source may be Group, Tape, Channel Input (pre-VCA fader) or Channel Output (post-VCA fader). If INPUT and OUTPUT are both selected, small fader source is pre-VCA but post any channel signal processing.

MIX BUS ROUTING is determined by a three position rotary switch. Panning is provided between the left and right inputs to the selected stereo mix buses. A master mix matrix panel provides further routing of the A, B and C stereo mix buses to the Programme Output.

CHANNEL SOLO & CUT functions are provided by large illuminated pushbuttons. Solo is normally In-Place, but may be switched to AFL (non-destructive) using the master AFL or Status Lock controls.

COMPUTER STATUS for each channel is normally set by the software in accordance with the particular mode of operation. The local Fader Status button is used to request special modes. Status is indicated by the red and green LEDs in various steady-state and flashing modes. Automatic nulling simplifies updates.

TOTAL RECALL™ automation of all I/O module switches and pots (except the Record button) is available as an option. The Status button is used to select individual channel displays in this mode.

VCA CONTROL GROUP SELECTORS allow the engineer to place any channels under the control of a centrally located Control Group Master. Position "I" isolates the fader from the master VCA Trim. Position "0" indicates local control. Positions 1-8 assign the channel to the selected Control Group.

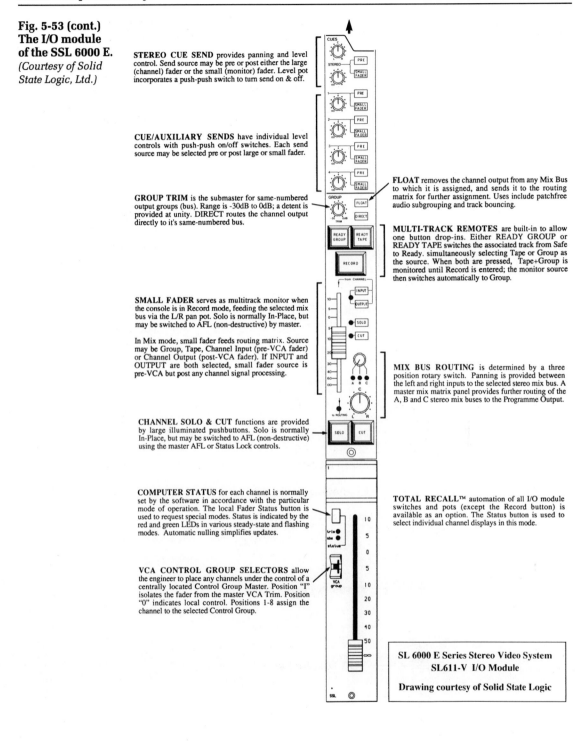

SL 6000 E Series Stereo Video System
SL611-V I/O Module

Drawing courtesy of Solid State Logic

Fig. 5-54. Sony/
MCI JH-618
GOVU I/O mod-
ule and stereo
line input mod-
ule. *(Courtesy of
Sony Corporation
of America, Inc.)*

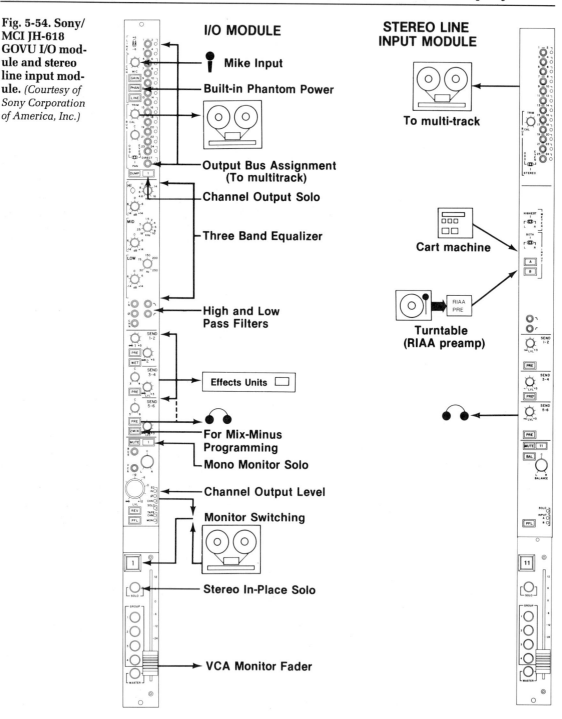

I/O MODULE

Mike Input

Built-in Phantom Power

Output Bus Assignment
(To multitrack)

Channel Output Solo

Three Band Equalizer

High and Low
Pass Filters

Effects Units

For Mix-Minus
Programming
Mono Monitor Solo

Channel Output Level

Monitor Switching

Stereo In-Place Solo

VCA Monitor Fader

**STEREO LINE
INPUT MODULE**

To multi-track

Cart machine

Turntable
(RIAA preamp)

The following sections describe the main features of the configuration of the standard I/O module.

The Microphone/Line Preamplifier

The first stage at the input of any audio production console is the mic/line preamp. Since the output of a microphone is often very low in level, a specially designed amplifier is required to raise it to the standard operating level (+4 dBm at 0 VU) of the console. Often an overall preamp signal-to-noise ratio in excess of 125 dBm is required in order to amplify the microphone signal to the proper level that will maintain the console's signal-to-noise ratio within a 70-dB dynamic range.

In order to compensate for the wide range of sensitivities of available microphone and line outputs and to optimize the system's signal-to-noise ratio, a preamplifier gain trim is incorporated as a means of controlling the gain of the preamp over an approximate range of 40 dB. This gain control may be designed in the form of a continuous trim pot or with a number of switchable steps.

The output of the preamp is often fed to a phase-reversing switch, allowing the operator to affect a 180° change of phase of the input signal.

Equalizers

Most production consoles provide facilities for some form of equalization (EQ), ranging from three frequency bands (including high- and low-band shelving with a selectable peak or dip as middle band) to a four-band equalizer with full parametric control over each band. Often low-pass and high-pass filters are present to take care of the attenuation of troublesome signals above or below the audio range with minimal or no effect on the sound of the program material. These frequency-selective filters may be switched in or out of circuit at will by the operator, with no audible effect.

Auxiliary Send Controls

The auxiliary send, or effects send, controls of an audio production console may be viewed as a highly flexible set of submixers. This part of the system may be organized in different ways for use with a large number of applications, but the most common are the following:

- As an effects side chain, allowing for overall input submixing to a direct or patched line feed to one or more effects devices (within the mixdown phase, this type of system allows for the effects devices to be returned and blended into the overall stereo output)

- As a versatile set of line output feeds that may be utilized for a variety of applications, including simultaneous monaural mix, musician cue feed, stage monitor mix, and live transmission feed

Auxiliary (aux) send controls of audio production consoles vary from a two-send capability to the eight-split stereo sends found on larger models. Reverb or echo sends are very often delegated to a separate stereo effects side chain.

Record Monitor Controls

Within the operating structure of the multitrack console, the record monitor controls often play a vital production role. This part of the system allows the sound engineer and the artist being recorded to monitor newly recorded and live studio audio in stereo and with a fully balanced perspective.

Configured as a subsection of the I/O strip, the record monitor controls consist of a stereo monitor submixer with 24 or more inputs. A stereo mix of the multitrack master tape can therefore be made during track building, recording, and overdubbing. This mix may be fed as a simultaneous stereo signal to the control room or to studio monitors, and often separate headphone (cue) balances may be given in either mono or stereo to the performers located within the studio. On the larger professional consoles, each independent I/O module will automatically derive the proper source signal for monitoring from either the multitrack ATR or the program input, depending on the operating status of the console, which can be selected as record, overdub (OD), or mixdown.

Channel Assignment Matrix

Distribution of any input signal may be made to any or all track inputs of the multitrack ATR through the use of the channel assignment matrix. This part of the I/O module provides ease and flexibility of channel assignment, doing away with the need for direct patching or normalization of the output signal. On the latest console designs, a facility for group panning has been added, allowing for the stereo routing of a combination of inputs to two audio tracks. One advantage of the group pan control is that it can be used to reduce the number of tracks needed for a mix by creating a stereo submix of a section of a musical or sound script.

The Master Output Module

The *master output module* operates as a complement to the I/O module in that it acts to receive all signals of the auxiliary sends as well as stereo mix information from each I/O module. It then forwards the overall signal and mixing levels to a master stereo output bus for distribution to an associated ATR (2-track or 4-track). The master output module often contains all of the necessary electronics for the final control and distribution of the stereo mixdown signal, including those for master auxiliary send levels, master auxiliary returns and pans, master output bus fader(s), console status selection, overall monitor trim and source selection, lineup oscillator, and talkback and slate facilities.

Digital Technology for Signal Routing

In parallel with recent advances in digital audio, major developments are occuring in the use of digital technology for signal routing within the modern audio production console. Through the use of a MOSFET (metal oxide semiconductor field-effect transistor) switching matrix and with the aid of a computer/keyboard interface, the majority of signal routing functions (which in the past were effected by means of hundreds of dedicated and potentially faulty switches) may be quickly and easily executed, memorized, reconfigured, and recalled. Total control over track assignment, auxiliary send and return routing, channel status, equalizer input and output, and external event structure is centrally assigned to create a multitude of flexible signal paths through the use of the system's keyboard.

Since all the above functions may be digitally encoded and stored in a short-term (scratch pad) or long-term computer memory, it is possible to recall all setup and routing information quickly. When operating in conjunction with automated mixing of levels and processors, automated signal routing allows for almost total recall of session and mixdown information within minutes, greatly simplifying the carrying out of complex signal-patching schemes.

Fig. 5-55 shows the Sony MX-P3036 automated recording/remixing console, which incorporates digital signal routing.

Fig. 5-55. Sony MX-P3036 automated recording/ remixing console. *(Courtesy of Sony Corporation of America, Inc.)*

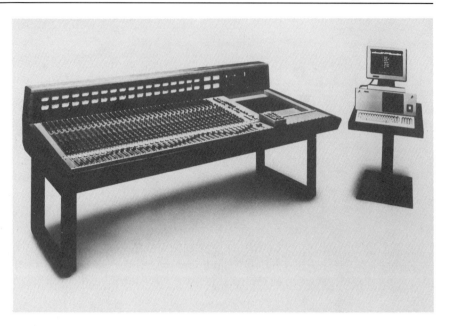

Automation of Level Balancing

The final mixdown phase for a video soundtrack may vary widely in complexity; anything from a simple mix consisting of a dialogue and presence track to a major production endeavor involving separate sound mixers for dialogue and SFX or music may be required. Given this potential complexity and the necessity of maintaining complete control over level, effects, reverberation, panning, and equalization on each audio track during the long periods of time required to mix a scene properly onto tape, it becomes obvious that the mixdown process could easily become overly burdensome. The answer to this problem is to employ a device that is capable of storing and automatically recreating any changes in level settings that are made by all of the associated faders used within a mix. This device must be flexible enough to allow the sound mixer to improve on these balances until the desired final mix is achieved. Such dynamic level balancing is made possible by a process known as *automated mixdown*, or *computer-assisted mixdown* (Fig. 5-56).

Currently, two types of systems exist for achieving overall automation of level balancing within an audio production console: the VCA (voltage-controlled amplifier) and the automated servo-driven fader. Both the VCA and the servo-driven system operate in conjunction with a program interface device for the proper encoding and decoding of associated fader levels and a volatile or nonvolatile medium for storage of this information.

The Voltage-Controlled Amplifier

The *voltage-controlled amplifier*, or *VCA* (Fig. 5-57), is employed within an automated mixdown application as an interface between the analog audio signal (whose level is to be stored and/or recalled) and a static measurable parameter (DC voltage) to be stored by the automated system's digital storage device (Fig. 5-58).

Fig. 5-56. The CASS-1 Computer-Assisted Sound Sweetener (shown here in the console automation mode) is able to simultaneously control up to 6 ATRs and 15 additional sources during audio editing, while also allowing for the automated mixing of an entire soundtrack. *(Courtesy of CMX, Inc.)*

With a VCA, the level of the program audio is a function of a DC voltage that is applied to the input control of the device. As this control voltage is increased (in relation to the position of the fader), the attenuation of the analog signal is likewise increased in direct proportion.

As well as controlling one VCA using a single control voltage, it is possible to control the level of the audio signals through more than one VCA in tandem using a single voltage control source. In this manner, any combination of independent audio signals may be ganged together and controlled from one voltage source (a fader). This technique is known as the creation of a signal group, or *grouping*.

Fig. 5-57. Three voltage-controlled amplifiers (VCAs) that employ the Allison Research EGC-101. *(Courtesy of Valley People, Inc.)*

Fig. 5-58. Master Mix MX644 automation computer. *(Courtesy of Audio Kinetics.)*

Modes of VCA Operation—An audio production console that is equipped with facilities for VCA automation is designed to operate in three basic modes: the write mode, the read mode, and the update mode.

In the *write mode*, the system's encoder will scan the DC-controlled input levels of each of the voltage-controlled input strips (as in-

Fig. 5-59. Representative VCA-based circuit for level control, employing both single control and grouping faders.

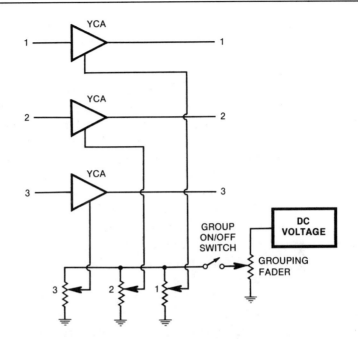

dicated by the console's fader mixing positions) in sequence at a high scan rate. Then the encoder digitally encodes this stepped control voltage information into memory. Depending on the system, this memory may be computer-based (synchronous diskette) or may consist of storage alongside the mix on two dedicated audio tracks of the multitrack tape (known as data tracks).

In the *read mode*, the system's decoder scans the computer memory or recorded data tracks and reconverts the digitally encoded information into equivalent DC voltages. These voltages are then fed to their respective VCAs or VCA groupings. In this manner, the elements of a complex soundtrack can be entirely reassembled under the automated control, and the final mix can be laid down to an ATR or layed back to the video master in a single first-generation pass, directly from the multitrack.

If a change in the level of program audio level needs to be made once the data track has been written, it is necessary to perform an update of the data track. At this point, any one or more automated tracks or channels may be switched into the *update mode*, thereby allowing for adjustment in levels to be made. In the update mode, manual matching of fader and data level is eliminated when punching into an existing data track. Settings may be changed by simply adding to or subtracting from a reference control voltage (with reference to a standard fader position) rather than by completely rewriting the information. In this way, an overall level correction may be made, without the need to reset all of the relative balances required to recreate complex track.

The Automated Servo-Driven Fader

The other means of achieving console automation is the *automated servo-driven fader*, which is marketed by Rupert Neve, Inc. and the Droid Works. The Neve computer-assisted mixing, or NECAM, system is designed for exclusive use in Neve consoles and operates by way of a touch-sensitive servo-driven fader. Unlike a VCA-based system, where DC voltage levels control the automation of levels, the NECAM fader is actually a resistive attenuator that is automatically driven by means of an interface with a servo motor. Thus, during a replay of an automated mix, the faders will move automatically in accordance with the requirements of each specific balance.

The NECAM 96 automation system does not make separate provisions for read, write, and update modes; it is billed as an instinctive automation system. All that is required with this system, once the desired rough mix has been achieved, is for the operator to move the fader that is to be updated to the desired location. The processing computer will then automatically store this information as updated material. The grouping of faders is assignable; there may be any number of groups, and a group may be moved in unison simply by touching any fader within that group. Mute functions, mute groups, and up to 128 event-switching functions may be called up by the control keyboard and stored within the microprocessor-based system. Automated data may be stored as a virtual mix (allowing an unlimited number of updates to be stored in a temporary memory), stored on disk drive (making a permanent record), or stored as any of 999 "snapshots" of fader and mute functions for instant recall.

The Audio Production Console in Video Broadcasting

Although the drive for quality is intense within the production facility for live broadcasting, the demands made on an audio production console there often differ widely from those imposed within a music recording facility or audio-for-video postproduction facility. In addition to the increased number of inputs and outputs required for video broadcasting, there are several other factors to consider. The console

Fig. 5-60. Production system layout with Sony JH-618 GOVU console. *(Courtesy of Sony Corporation of America, Inc.)*

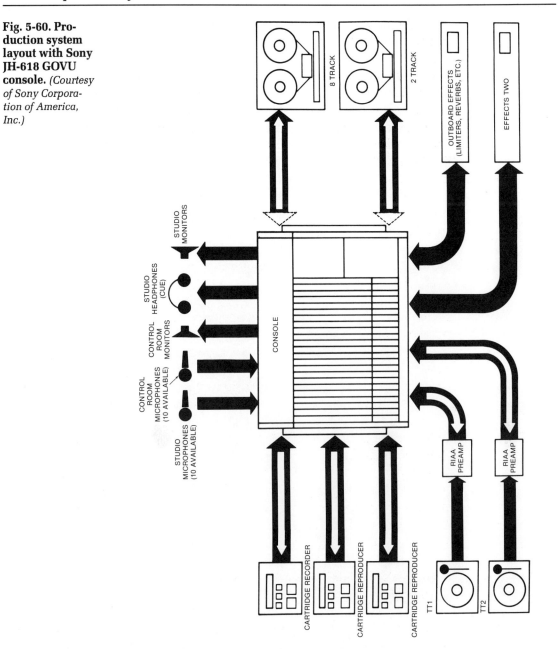

must be able to easily control its various inputs. It must be reliable, rugged, and relatively easy to learn to operate. The console must also be fast, since time is money (and often reputation) within the "on the air" environment. Thus, it must be ergonomically designed with controls that have a logical order. Within stereo production, the employment of ganged combinations of inputs and faders makes program control easier and more straightforward. Automatic event switching to start a cartridge machine, phono, etc. is also often a necessary feature in live production.

6 *Introduction to Electronic Editing Techniques*

In video production, it is indeed the exception when raw program material is used as is from beginning to end. More often, program material is pieced together from related but separate scenes, including videotape footage and recorded sound sources, in order to create the final master tape. This method of production is known as the editing process.

Videotape Editing

The piecing together of separate sections of analog audio tape (and certain forms of longitudinal digital audio tape) is a relatively simple matter. A physical splice (tape cut) is made at the appropriate points on the tape, and the two program pieces are joined together with splicing tape. As this splice passes the audio playback head, the signals within the transition are averaged out (either magnetically or electronically), resulting in a clean and continuous audio signal. Such a physical edit is often undetectable, even to the professionally trained ear.

In editing videotape, an entirely different approach must be taken to join together separate pieces of program material. During the early years of video production, the use of the quad VTR allowed for the physical cutting and splicing of program material; the video tracks for a quad machine make an angle of almost exactly 90° with the longitudinal dimension of the tape. With this method, the video material to be spliced was located with the help of a mechanical tape footage counter, or even a microscope. The edit points (located between frames within the vertical interval) were found, and the tape was cut with a razor blade and then spliced with another piece. Since there was no way to preview an edit, the physical editing of videotape was rather clumsy, occasionally causing picture breakup or rollover at the splice point.

Physically cutting videotape for a helical scan VTR would be virtually impossible. This is due to the small angle at which the video track cuts across the tape width. A cut would have to follow the precise video track angle; otherwise, the result would be a bar of noise at every field represented within the edit point. Even if a perfect cut were made, the required splice joint would be impossible to execute. From this, it is obvious that some method other than old-fashioned splicing is required in order to perform video edits within the modern production facility. Since the videotape cannot be physically cut in order to perform the editing process, *electronic editing* must be employed.

Early techniques of electronic editing (first developed in 1961) were generally unreliable and often resulted in picture breakup or inaccurate cut-in placement, a phenomenon that earned the rather unloving nickname of "edi-crash." However, developments soon allowed for a smooth transition from one video signal to another on a single, continuous piece of videotape.

The process of electronic editing must fulfill two basic conditions in order to be effective and useful:

1. The quality of the final edited master videotape must be equal to that of the source tape(s).
2. Switching between various video sources must be both easy to accomplish and undetectable in the final product.

These conditions allow for the transfer of various, separate video source signals onto a final videotape, which is known as the *production master*, or *edited master (EM)*, and contains a continuous program consisting of quality video from beginning to end. Original program material is rerecorded onto the EM from various sources, in a piece-by-piece fashion, until the final edited video program has been assembled.

Servo Control During the Video Edit

The central concept behind the process of electronic editing is signal timing when switching from one video source to another. The stability of a video signal is extremely critical for accurate image reproduction. This factor is of equal importance during the video edit.

In order for a video edit to take place, the timing of the video signal that is already recorded on the videotape must match precisely with that of the incoming video signal that is to be edited into the program (Fig. 6-1). When this requirement of vertical interval alignment is met, the transition from playback to record within the VTR will be made without distortion or breakup of the picture. Since the incoming video signal to be recorded is generated in real time, the only way to lock it to the VTR's reproduced signal is to exercise control over the tape transport itself. Precise control over tape motion allows the reproduced signal to be brought into exact vertical interval alignment with the incoming signal. This alignment is achieved by locating the control track on the tape and comparing it with the incoming vertical sync pulse. The difference in pulse location is determined by a basic resolver, and a correcting voltage is sent to the VTR's capstan servo until the two signals have been brought into synchronization. Once the tape and incoming signals match up, the servo that drives the rotating video head must be brought into sync with the reproduced vertical interval (Fig. 6-2). Once this has been accomplished, the recorded and incoming signals will be in step with each other, and the VTR can be switched from the reproduce mode into the record mode. Upon playback of the newly edited material, the video and control track signals will match up at the edit point, causing no visible picture distortion.

Switching Between Video Signals

Once the recorded and incoming video signals have been brought into synchronous lock, two methods may be employed within a video edit for switching between the signals: random switching and vertical interval switching.

In *random switching*, the incoming video signal and an erase current are applied to the appropriate heads as soon as the record button is depressed. This method allows the switching between signals to occur at any point, even within the visible picture area. Random switching is easily accomplished using a minimum of electronics and is used with considerable success with nonbroadcast-type VTRs.

With *vertical interval switching*, however, the signal and erase current are not applied to the appropriate heads at the instant the record button is depressed. Instead, the transport is placed into the record mode when the next vertical interval on the tape has been reached. The

Fig. 6-1. Align-
ment of video
signals during
the video edit.

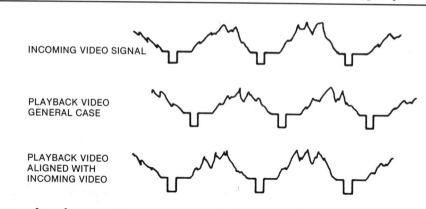

INCOMING VIDEO SIGNAL

PLAYBACK VIDEO
GENERAL CASE

PLAYBACK VIDEO
ALIGNED WITH
INCOMING VIDEO

record and erase currents are applied within this interval, outside of
the visible field (Fig. 6-3). Vertical interval switching is currently the
broadcast industry standard.

Fig. 6-2. Coinci-
dence of the
head and the re-
corded vertical
interval on the
control track.

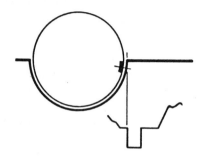

The Video Edit

Two types of edits are used for the placement of newly recorded material alongside a prerecorded video signal: the assemble edit and the insert edit.

The Assemble Edit

The simplest type of edit employed in video production is the *assemble edit*. One switchover from the reproduce mode to the record

Fig. 6-3. Switch-
ing occurs
within the verti-
cal interval,
away from the
visible picture
area.

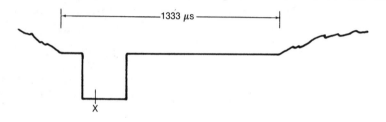

1333 μs

X

mode is made (Fig. 6-4). Thus, new material is placed on the tape and continues in an uninterrupted fashion, with no return to the previously recorded signal. This new material may originate from any stable video source, such as a live camera or a second VTR playing original field tape. The assemble edit is the most commonly encountered type in modern video production.

Fig. 6-4. With the assemble edit, the VTR is switched into the record mode within the next available vertical interval, recording new video and control tracks from the cut-in point onward.

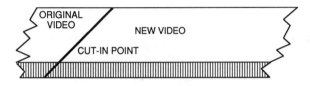

When an assemble edit is to be performed using a second VTR as the source of the new program material, both VTRs must first be placed into the reproduce mode, and time must be allowed for the record VTR to lock into sync with the control track pulses of the source VTR. Once this has occurred, the record mode may be entered (either by actually depressing the record button or by initiating a remote control function) and the new signal transferred to the record VTR. Thereafter, the new signal is treated just like the original video, with all selectable erase functions (main, control track, and audio) activated. Resolution between control tracks of the source and record VTRs is no longer necessary once the record mode has been entered. A new control track is put onto the tape. It will reproduce in perfect step with the control track of the original video signal, allowing for a smooth transition between original and new video images without rolling, jumping, or visible distortion.

The assemble edit may be a single program transition or have a duration of only a few frames (such as for animation production). The cut-out point is executed by initiating the stop mode. New material is built up in such a way that the appearance is of one complete and continuous program.

The Insert Edit

With the *insert edit*, the VTR switches from the reproduction of video material already on tape to recording incoming video signals, as with the assemble edit. However, with the insert edit, the transport may be returned to the reproduce mode at any time, leaving the original video signal intact following the cut-out point (Fig. 6-5). A segment of new material may thus be inserted into an existing program without visible breakup or distortion of the picture at the cut-in and cut-out points.

Fig. 6-5. With the insert edit, servo control is maintained in the reproduce mode, allowing synchronization with the original video material to be maintained so that stable cut-in and cut-out points are executed.

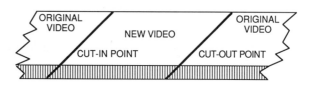

In the assemble edit, once the cut-in point has been reached, control track synchronization is no longer required, since there is no return to the original video signal. However, in the insert edit, this original signal is returned to at the cut-out point, so a greater degree of control over the transport's functioning is necessary. In performing an insert edit using a second VTR as the source for the new video material, both transports are initially placed into the reproduce mode and allowed to become synchronized. Once this has been accomplished, the record VTR may be placed into the record mode. At no point within an insert edit may control track synchronization between the source and record VTRs be stopped because the incoming and original video signals (and thus the control tracks) must be in precise alignment at the cut-out point if this transition is to occur without distortion or breakup of the video image. This means that the original video signal's control track must be left intact (must not be erased) and that it must act as the synchronous locking pulse for the incoming signal. At the cut-out point, the current carrying the incoming signal to the head may be switched off within the common vertical interval and the transport returned to the reproduce mode.

Auto-Framing

A feature found only on broadcast-quality VTRs is *auto-framing*. This feature ensures that the switch within a video edit occurs at the beginning of a color frame sequence and does not accidentally break up the A-B-A-B field pattern necessary for color reproduction. Thus, a correct matchup of color fields is checked for before the switch may take place within a vertical interval.

The Erase Function Within the Video Edit

At this point, it is worth considering the erase function within the video edit. Since the transition to the incoming signal must take place within a single vertical interval once the record mode is entered, the main video erase head will be ineffective for a short period of time. A gap representing up to 3 seconds of tape time may exist between this head and the rotating video record head, producing a section of un-erased video at the cut-in point (Fig. 6-6).

Fig. 6-6. The physical distance between the main erase head and the rotating record head in a VTR.

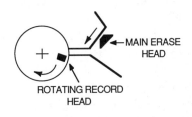

MAIN ERASE HEAD

ROTATING RECORD HEAD

In an assemble edit, it is necessary to erase this short section of tape until the main erase head is able to take over. In an insert edit, it is not possible to activate the main video erase head because this would produce a signal gap at the cut-out point. Some means of erasure must therefore be employed at the rotating head wheel itself, and it must be able to track the video signal precisely. Two such means of achieving erasure are an increase in the head current and the use of a flying erase head.

It is possible to erase a previously recorded signal by increasing the current supplied to the video head to 125% of conventional current levels. This serves to effectively obliterate any residual signals that exist on the tape, but an accompanying disadvantage is a resultant saturation of the recorded video signal. If the edit has to be reexecuted, increased head current may not be enough to obliterate this saturated signal, severely limiting production flexibility. Also, this type of erasure is unsuitable for performing insert edits.

The second type of erasure, commonly in use with broadcast-quality VTRs, is that utilizing the *flying erase head*. This device is simply a second erase head that is located on the head wheel itself, positioned ahead of the record/reproduce head. With this setup, individual video tracks may be erased within an assemble edit beginning at the cut-in point but may also be easily maintained throughout the entire execution of an insert edit. Using this type of erasure system, a video edit may be reexecuted until a satisfactory product has been achieved, which constitutes a high degree of production flexibility.

Electronic Video Editing

Although a great deal of complicated circuitry is required in order to perform video editing, execution of the process is quite simple in actuality. However, for professional productions, it is necessary that the timing of edits be highly accurate. The final frames of the original video material must blend smoothly into the beginning frames of the new material. Since both the record VTR and the source VTR must be rolling in order to accomplish an edit, this matchup is very difficult to achieve manually, or "on-the-fly." Therefore, if the editing process is to be creatively feasible and cost-effective, some form of automated control over the involved transports and their electronic functions must be employed.

In 1963, the Ampex Corporation developed the Editec, which expanded video editing capabilities by directly controlling the edit points electronically. Editec permitted the shifting of two edit points by a large number of picture frames and gave the operator the ability to preview an edit before actually executing it. This system operated by placing a beep tone onto the cue track of the record VTR (a quad VTR) at the desired edit cut-in point. When the edit was to be executed, both record and source VTRs were rolled back and the edit was electronically performed. The Editec system was the state-of-the-art model of its day, but it was slow, cumbersome, and did not perform with frame accuracy. Although the Editec system relied on the positioning of the control track signal for achieving relative video sync between the record and source VTRs, it used the "cue beep" to locate and execute the edit. Preroll location and basic transport functions (with the exception of record) had to be performed manually.

The Control Track Edit

The next generation of automated electronic editing techniques after Editec involved the *control track edit*. With this technique, the control track serves two functions: (1) video signal synchronization, and (2) relative videotape location on record and source VTRs to control edit timing. For a record/source VTR configuration, control track editing is still in widespread use throughout the world and is an effective and relatively inexpensive way to accomplish the basic edit function.

The *video edit controller* (Fig. 6-7) serves as the core of a control track editing system. All necessary control track and transport functions of the VTRs are terminated at this controller, which allows all major operational functions to be remotely controlled from its editor panel. The function of the edit controller is to monitor the status and tape location of both record and source VTRs and to exert a degree of transport and switching control over these devices. This control is maintained by electronically counting the control track pulses of the VTRs and keeping track of the relative transport positions by means of these counts. In operation, the cut-out and cut-in points for the edit may be automatically located on the videotape. This is accomplished through the use of search control dials located on the controller, allowing for full jog capabilities.

Fig. 6-7. The JVC RM-G85OU offline video edit controller with SMPTE and control track edit capabilities. *(Courtesy of JVC Company of America.)*

Once the edit points are located for both VTRs, they can be electronically marked by resetting the control track counters to zero. At this point, the automatic preroll function is engaged, and the edit controller will reverse wind each VTR to a preroll location 5 seconds in advance of the edit location. When the controller's edit or record button is depressed, both transports are automatically placed into the reproduce mode and brought into video sync. When countdown reaches the zero mark, the record VTR is switched into the record mode, thus beginning the assemble edit. Multiple edit points can often be established with the more advanced edit controller, allowing them to be used for insert edits and semiautomated assembly edits with a high degree of success.

The major drawbacks encountered with this type of editing control are the general lack of repeatability and the inability to make relative time (offset) adjustments. The control track edit may be regarded as being accurate if the task at hand is basically simple in nature and is given immediate production attention. As with the use of a simple resolver for synchronization (discussed in Chapter 3), the use of the control track as a means of edit control is relative. A control track edit controller is able to keep track of two videotape locations relative to one another; however, it is unable to determine exact locations. If this locked-in reference is interfered with (videotape taken off the transport, misreading of control track, etc.), all calculations for the edit will be referenced to video frames that are no longer accurate.

Repeatability and absolute control over tape position and offset adjustments are critical in professional video production. Therefore, the control track edit is generally utilized only in smaller facilities that cannot afford more sophisticated equipment. This form of editing is in regular use for ENG, industrial, and documentary productions, where financial constraints or deadlines are tight.

Editing with Time Code

The limited repeatability and the lack of accurate offset control or extensive automated control gradually made the relative control track edit ineffective for high-level video production. In 1967, a new method of performing video edits electronically was developed; it took advantage of the then newly standardized SMPTE/EBU time code. Using time code in the performance of the editing process brought about an unprecedented degree of production accuracy and flexibility. Some of the major advantages of editing with time code are as follows:

- Nonslip time code addresses, assuring absolute repeatability
- Frame-accurate editing capabilities
- Ease of automated system control
- Ease of audio-for-video production (audio building in the video edit, etc.)

At first, time code served as a basic means for locating specific cut-in and cut-out points for video edits. It also provided for an improved preroll function and frame accuracy in the switching between video signals. In this sense, the new method produced results similar to those achieved with a high-resolution control track edit controller.

On-Line Video Editing

The use of time code gave the process of video editing a higher resolution as well as the added benefit of frame-accurate repeatability. Accurate identification of frame locations on videotape has since produced an increased degree of automated control over the entire editing process. In 1972, computerized control systems were first introduced into professional video production facilities and they have developed since then into today's standard technology (Fig. 6-8). The use of computer-

Fig. 6-8. The on-line video edit suite.
(Courtesy of the Editel Group, Chicago.)

assisted tools for editing is commonly referred to as *on-line video editing.*

The function of an on-line video edit controller is to exercise interactive control over all devices within the video edit suite, so that operators can easily and effectively produce edited master videotapes. Simply stated, the video edit controller allows the operator to select material from video and audio source footage and electronically transfer it to master videotape or to an ATR or multitrack ATR. The controller keeps track of all information pertaining to each edit, saving this data for future reference.

As is illustrated in Fig. 6-9, the modern on-line video edit controller may be used to control a large number of synchronized devices. Such a video editing system consists of a varied array of equipment, including VTRs, VCRs, ATRs, video effects switchers, digital effects generators, character generators, matte cameras, etc. Each of these devices has its own operating requirements and characteristics. Using an approach called distributed processing, the controller integrates this wide range of components into an efficient system, thereby simplifying and speeding up the editing process.

At the completion of an editing session, the video edit controller produces three final products:

1. A printout (hard copy) of all the edit decisions that went into making up the program, including any additional notes or comments entered by the editor

2. A floppy disk or punched paper tape containing the same summary of edit decisions, which may be loaded into any EDL-compatible system, allowing for later program changes and for final automated assembly of the program

3. A master videotape of the edited program incorporating frame-accurate, stable edits

It is important to remember that an edit controller is a control device and not a processor or a switching device. This means that it is capable of proprietary control over the basic functions of a wide variety of devices specific to different media. An on-line edit controller may be equally at home in video production and audio for video, whether the media involved are analog or digital in nature.

It is equally important to remember that all VTRs, TBCs, time code generators, video edit controllers, and ancillary equipment must always be locked in video synchronization with a house sync pulse generator.

Fig. 6-9. Typical system connections for the Sony BVE-5000/ 5000P on-line video edit controller.
(Courtesy of Sony Corporation of America, Inc.)

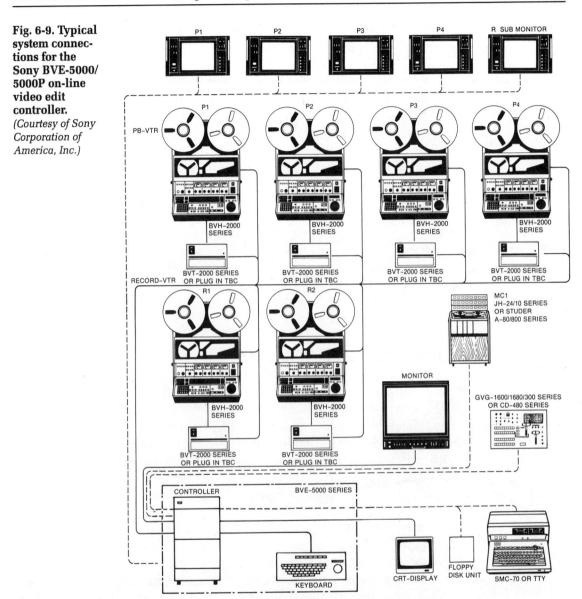

The following list gives a brief description of the components commonly found within an EDL-based video edit controller:

- The functions of the *central controller* (which consists of a microcomputer, a data entry keyboard, and a CRT monitor), are to process information inputted by the operator, to supervise communications among all involved pieces of equipment, and to coordinate the varied activities that take place during an editing session. This device operates in an interactive mode, maintaining a constant dialogue with the operator by displaying questions and messages on the CRT screen, and thus enabling the process to run smoothly.

- Each device is connected with the central controller by means of an *intelligent interface.* The central controller tells each device what to do, but it is the interface's role to figure out how its particular device is to get the job done. Thus, an intelligent interface is essentially a microprocessor that is specifically programmed for controlling the unique characteristics of one type or model of VTR, VCR, ATR, or effects device.

- A *line printer* is employed so the video edit controller can produce printed copies of the EDL.

- The *floppy disk drive* allows the operator to load the system operating program (where applicable) and edit information into the central controller's memory. It also provides storage for EDL files.

The CMX 3400 Video Edit Controller

The CMX 3400 (Fig. 6-10) is a computer-based video edit controller designed to replace the industry standard CMX 340X on-line system. The 3400 is capable of maintaining control over an array of up to 24 VTRs, ATRs, effects devices, and other devices during automated edits. This software-based modular system maintains a high degree of control over functions and switching through the use of the standardized CMX edit decision list, or EDL.

The Sony BVE-5000 Video Edit Controller

Designed for both on-line and off-line applications, the Sony BVE-5000 (Fig. 6-11) is a versatile video edit controller. It utilizes the CMX-compatible edit decision list, which is available on punched paper tape, on 8-inch floppy disk, and in printed form. The BVE-5000 is able to control a maximum of six source VTRs and two record VTRs, with a memory capacity for up to 512 individual events on the EDL (the total number of events is proportionally reduced if the event sequence contains special effects). A full range of constant and/or variable events, single "split events" or multiple events, and audio advances or delays may be programmed into the EDL format from the system keyboard. Optional interfaces that allow Sony/MCI and Studer ATRs to be added to the system serve to simplify the integration of au-

**Fig. 6-10. The
CMX 3400 video
edit controller.**
*(Courtesy of
CMX, Inc.)*

dio production facilities into the video editing suite. One unique feature of this controller is the storage of the system's operating program in read-only memory (ROM), thus avoiding the need for manual loading prior to an editing session.

**Fig. 6-11. The
Sony BVE-5000
video edit
controller.**
*(Courtesy of Sony
Corporation of
America, Inc.)*

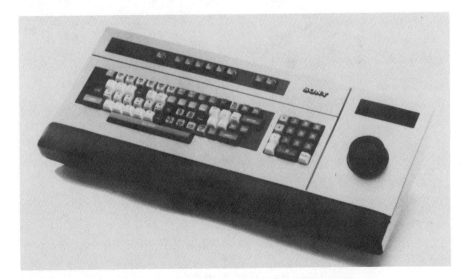

The Adams-Smith 2600 A/V Edit Controller

The Adams-Smith 2600 A/V edit controller (Fig. 6-12) provides both complete EDL-based video editing facilities and the unique capability of fully integrating the audio sweetening process into the video edit.

The 2600 A/V may operate as an on-line or off-line video editor, an audio-for-video laydown editor, a double-system audio editor/synchronizer, and an EDL-based controller for audio sweetening. All of this is accomplished by providing intelligent interfaces for synchronizing the ATRs. These interface modules simulate the dynamic functioning and controls of VTRs. Thus, the edit controller itself doesn't "know" what the interfaced transports are. One feature of the 2600 A/V is the clutch/declutch key, which allows any pair of sources to operate as a synchronized source. Since the 2600 A/V can control up to ten transports, a double-system configuration could consist of four VTR/ATR pairs as sources with a VTR/ATR recording combination operating in clutched tandem.

Fig. 6-12.
Adams-Smith
2600 A/V edit
controller.
(Courtesy of
Adams-Smith.)

The Edit Screen

The edit decision list developed by CMX, Inc., is generally accepted as the standard in the video industry for performing automated edits. Other manufacturers (such as Sony, Adams-Smith, Audio Kinetics, etc.) have developed CMX-compatible operating programs for use in their video and audio-for-video edit controllers.

The following is a detailed description of the edit decision list as it is currently used by the CMX 3400 video edit controller.

The EDL screen display contains three main areas (Fig. 6-13):

1. Menu area
2. System message area
3. Edit decision list area

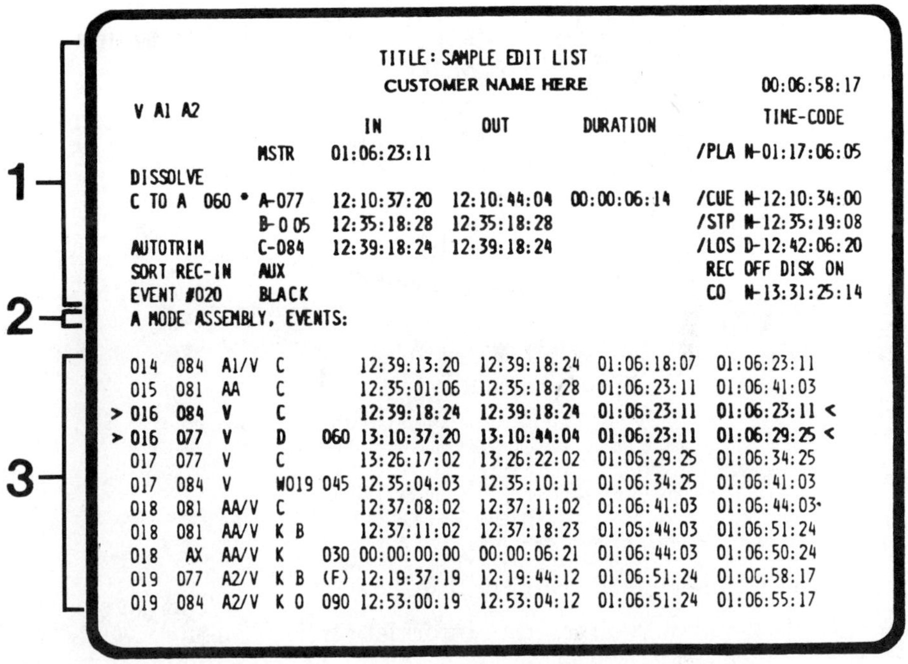

Fig. 6-13. The CMX 3400 edit screen.
(Courtesy of CMX, Inc.)

```
                              TITLE: SAMPLE EDIT LIST
                              CUSTOMER NAME HERE                    00:06:58:17
        V A1 A2
                                  IN         OUT      DURATION       TIME-CODE
                  MSTR      01:06:23:11                             /PLA N-01:17:06:05
        DISSOLVE
        C TO A 060 •  A-077  12:10:37:20  12:10:44:04  00:00:06:14  /CUE N-12:10:34:00
                      B-0 05 12:35:18:28  12:35:18:28               /STP N-12:35:19:08
        AUTOTRIM      C-084  12:39:18:24  12:39:18:24               /LOS D-12:42:06:20
        SORT REC-IN   AUX                                           REC OFF DISK ON
        EVENT #020    BLACK                                         CO  N-13:31:25:14
        A MODE ASSEMBLY, EVENTS:

        014  084  A1/V  C        12:39:13:20  12:39:18:24  01:06:18:07  01:06:23:11
        015  081  AA    C        12:35:01:06  12:35:18:28  01:06:23:11  01:06:41:03
      > 016  084  V     C        12:39:18:24  12:39:18:24  01:06:23:11  01:06:23:11 <
      > 016  077  V     D    060 13:10:37:20  13:10:44:04  01:06:23:11  01:06:29:25 <
        017  077  V     C        13:26:17:02  13:26:22:02  01:06:29:25  01:06:34:25
        017  084  V     W019 045 12:35:04:03  12:35:10:11  01:06:34:25  01:06:41:03
        018  081  AA/V  C        12:37:08:02  12:37:11:02  01:06:41:03  01:06:44:03•
        018  081  AA/V  K B      12:37:11:02  12:37:18:23  01:05:44:03  01:06:51:24
        018  AX   AA/V  K    030 00:00:00:00  00:00:06:21  01:06:44:03  01:06:50:24
        019  077  A2/V  K B  (F) 12:19:37:19  12:19:44:12  01:06:51:24  01:0C:58:17
        019  084  A2/V  K O  090 12:53:00:19  12:53:04:12  01:06:51:24  01:06:55:17
```

The Menu Area

The *menu area* displays information pertaining to the edit file currently being set up. The menu area is divided into four sections (Fig. 6-14):

1. Function display
2. Assigned channel display
3. Edit point display
4. Current status display

The *function display* gives five types of information:

1. A/V select mode—a total of eight possible alternatives
   a. Audio and video—three choices
      (1) V   A1
      (2) V   A2

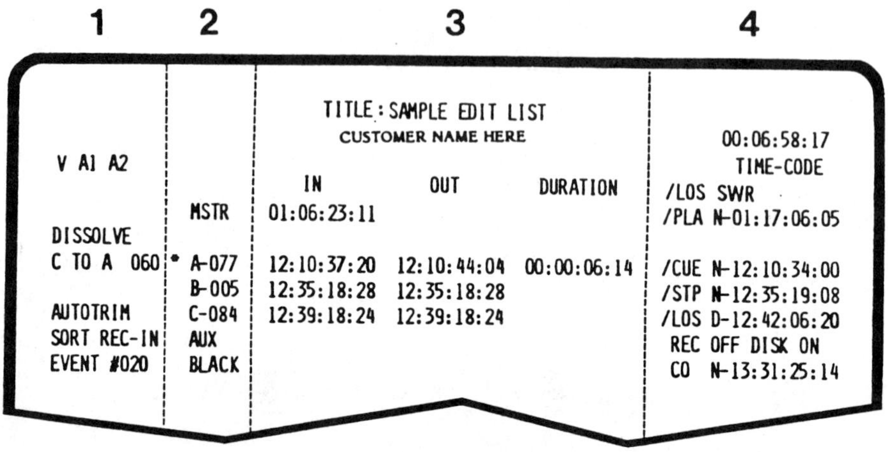

Fig. 6-14. The menu area of the CMX 3400 edit screen.
(Courtesy of CMX, Inc.)

```
        1        2              3                    4

                        TITLE: SAMPLE EDIT LIST
                          CUSTOMER NAME HERE              00:06:58:17
   V A1 A2                                               TIME-CODE
                         IN        OUT      DURATION   /LOS SWR
                  MSTR  01:06:23:11                    /PLA N-01:17:06:05
  DISSOLVE
  C TO A 060 * A-077  12:10:37:20 12:10:44:04 00:00:06:14  /CUE N-12:10:34:00
              B-005   12:35:18:28 12:35:18:28          /STP N-12:35:19:08
  AUTOTRIM    C-084   12:39:18:24 12:39:18:24          /LOS D-12:42:06:20
  SORT REC-IN AUX                                      REC OFF DISK ON
  EVENT #020  BLACK                                    CO  N-13:31:25:14
```

       (3) V   A1   A2

    b. Video only (V)

    c. Audio only—three choices

       (1) A1

       (2)   A2

       (3) A1   A2

    d. Split edit (A/V)

2. Transition information (two lines)

    a. Type of transition—four possible alternatives

       (1) CUT

       (2) DISSOLVE

       (3) WIPE (with pattern number)

       (4) KEY or KEY OUT

    b. From and to sources

    c. Rate of transition in frames

3. AUTOTRIM or AUTOCLEAN (either is displayed when that function is activated)

4. Sort mode—three possible alternatives

    a. SORT OFF

    b. SORT REC-IN

    c. SORT EVNT #

5. Event number (number of current edit)

The *assigned channel display* shows the master (MSTR) and source(s) that have been assigned during the initialization procedures, plus AUX and BLACK. These are all the channels that will be used during the edit session. Next to each source is a three-digit reel number, also assigned during initialization. An asterisk (*) will be displayed before one of the assigned channels, indicating that it is the active channel, that is, that commands may be given to it.

The *edit point display* displays the in and out times (under the headings IN and OUT) and the durations that have been entered for the assigned channels. Above the edit point display is the show title, which is assigned during initialization. Directly below the title appears the customer's name, which is recorded onto the disk at the CMX factory.

The *current status display*, which is at the far right of the menu area, presents information pertaining to the present status of each interfaced device and the system mode. This display includes:

- VTR mode
- Time code mode
- Machine time code
- Other current status data

The System Message Area

The *system message area* (Fig. 6-15) is the second main area of the edit screen and appears in the middle portion of the display below the menu area. The system message area is where the major portion of the dialogue between operator and system takes place. The information displayed in this area may be broken down into the following categories:

- Data entry functions
- System status messages
- Informal messages
- Error messages

Any or all of the above messages and questions will appear in the system message area.

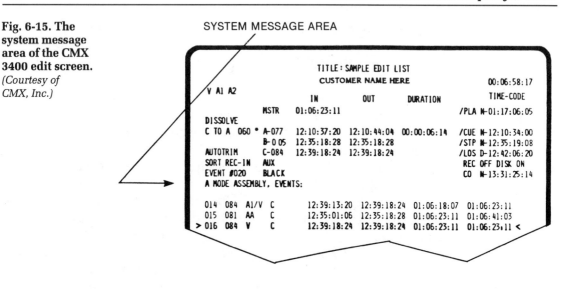

Fig. 6-15. The system message area of the CMX 3400 edit screen. *(Courtesy of CMX, Inc.)*

The Edit Decision List Area

The third main area of the edit screen is the *edit decision list area*. All edits that have been entered into the computer by the operator are displayed in a list for easy reference. The edit that appears brightest is called the *highlighted event*. The display that appears in the edit decision list area can also be printed out as hard copy and will appear exactly as it does on the screen except there will be no indication of drop frame and nondrop mode.

In Fig. 6-16, the columns have been numbered across the top of the edit decision list. The following is an explanation of the contents of each column:

1. *Event number*—Each three-digit entry in the first column is the event number of an edit.

2. *Reel number*—Each three-digit entry in the second column is the reel number for the source associated with the given edit. This number tells which reel contains the proper source material. Other than reel numbers, two other symbols may appear in this column: BL stands for black, and AX stands for an auxiliary source.

3. *A/V mode*—The third column tells the operator which audio and/or video mode is used for the specific edit. Note that these indicators are more condensed than those in the function display. The possible codes appearing in this column are as follows:

A = audio 1
A2 = audio 2
AA = audio 1 and audio 2
AA/V = audio 1, audio 2, and video
B = audio 1 and video
A2/V = audio 2 and video
V = video only

4. *Transition type*—The fourth column indicates which type of transition is used within the edit. The following entries may appear in this column:

C = cut
D = dissolve
W = wipe (a three-digit number indicates the code number of the wipe pattern)
K = Key-in edit
KO = Key-out edit
KB = Key-in background source

5. *Transition rate*—The three-digit number in the fifth column indicates the rate of transition in frames. An entry of 000 or no entry indicates that the edit is a direct cut. Dissolves, wipes, and key-ins are always accompanied by an entry in this column. Another possible entry here is (F), which is used with key-ins that have both sources fading either to or from black.

6. *Playback source in-time*—The sixth column displays the in-time, or edit in-point, for a scene. This time code location gives the first frame that will be seen in a video edit.

7. *Playback source out-time*—The seventh column displays the out-time, or edit out-point, for a scene. In-times and out-time of all scenes are pinpointed by these code locations in the sixth and seventh columns, so it would be easy to locate the original scene material if changes had to be made within the edit.

8. *Record master in-time*—The time code address in the eighth column indicates exactly where on the master videotape a specific scene has been laid down.

9. *Record master out-time or duration*—The time code address in the ninth column tells the operator where the scene ends on the master videotape. The 3400 can instead display the total duration of an edit instead of the out-time if the operator switches the OUT/DUR toggle.

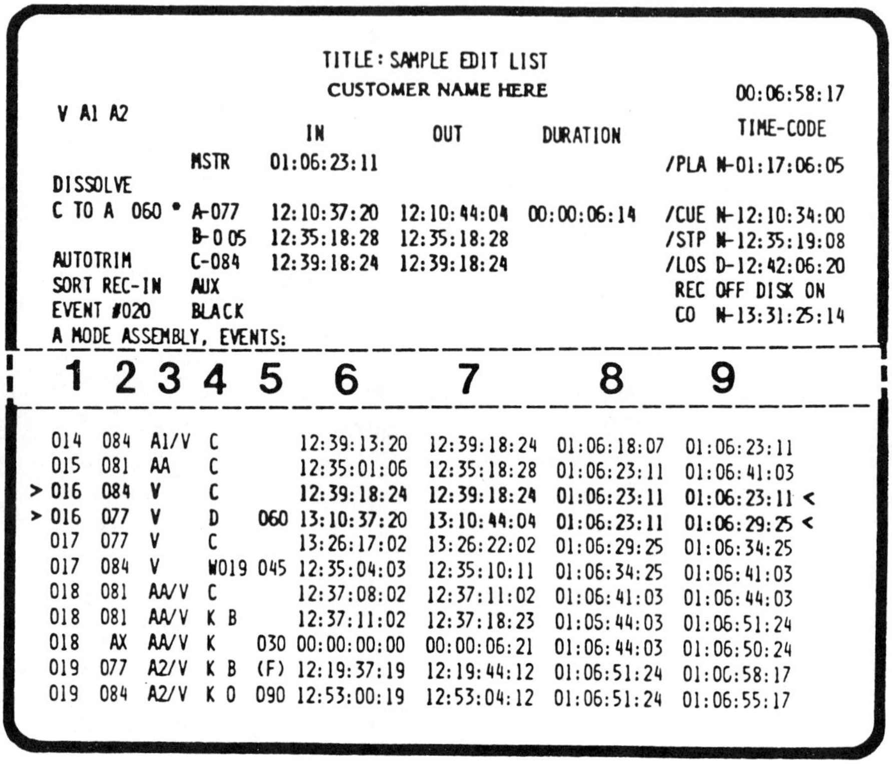

Fig. 6-16. The
CMX 3400 edit
decision list
area.
*(Courtesy of
CMX, Inc.)*

Figure shows CMX 3400 edit decision list display:

```
                        TITLE: SAMPLE EDIT LIST
                        CUSTOMER NAME HERE                    00:06:58:17
        V A1 A2
                              IN        OUT       DURATION       TIME-CODE
                MSTR   01:06:23:11                          /PLA N-01:17:06:05
 DISSOLVE
 C TO A  060 •  A-077  12:10:37:20  12:10:44:04  00:00:06:14 /CUE N-12:10:34:00
                B-0 05 12:35:18:28  12:35:18:28             /STP N-12:35:19:08
 AUTOTRIM       C-084  12:39:18:24  12:39:18:24             /LOS D-12:42:06:20
 SORT REC-IN    AUX                                          REC OFF DISK ON
 EVENT #020     BLACK                                        CO  N-13:31:25:14
 A MODE ASSEMBLY, EVENTS:

    1   2  3  4  5    6       7         8        9

   014  084  A1/V  C       12:39:13:20  12:39:18:24  01:06:18:07  01:06:23:11
   015  081  AA    C       12:35:01:06  12:35:18:28  01:06:23:11  01:06:41:03
 > 016  084  V     C       12:39:18:24  12:39:18:24  01:06:23:11  01:06:23:11 <
 > 016  077  V     D   060 13:10:37:20  13:10:44:04  01:06:23:11  01:06:29:25 <
   017  077  V     C       13:26:17:02  13:26:22:02  01:06:29:25  01:06:34:25
   017  084  V     W019 045 12:35:04:03  12:35:10:11  01:06:34:25  01:06:41:03
   018  081  AA/V  C       12:37:08:02  12:37:11:02  01:06:41:03  01:06:44:03
   018  081  AA/V  K B     12:37:11:02  12:37:18:23  01:06:44:03  01:06:51:24
   018  AX   AA/V  K   030 00:00:00:00  00:00:06:21  01:06:44:03  01:06:50:24
   019  077  A2/V  K B (F) 12:19:37:19  12:19:44:12  01:06:51:24  01:06:58:17
   019  084  A2/V  K 0 090 12:53:00:19  12:53:04:12  01:06:51:24  01:06:55:17
```

Indicators of Drop Frame and Nondrop Mode—On the screen display
of the 3400's edit decision list, the operator is able to distinguish between drop frame and nondrop mode for the various reels according to
whether a semicolon (;) or a colon (:) is used between the time code address numbers. These indicators appear only on the edit decision list
displayed on the 3400 screen. The hard copy printout of the list will
display only colons in order to ensure compatibility with other EDL-
based systems.

Encoding Videotapes with Time Code

Computer-assisted electronic editing is made possible through the use
of SMPTE/EBU time code. Without time code addresses, there would
be no means of accurately identifying each frame of video and/or the
corresponding audio on associated field and master tapes during
postproduction. Also, event-programmable devices may be triggered
through the use of time code commands.

When employing time code within a computer-based editing system, it is important that each roll of videotape that is to be used as an edited master be prepared in advance of the actual edit session. For current 1-inch, type-C formulations, the on-line edit facility must provide the client with a prestripped *black tape* containing video black, LTC and/or VITC time code, and audio silence.

Black burst is used for recording a control track onto a videotape prior to the edit session in order to attain video sync with the source VTR(s). During the process of editing, the black signal will be replaced frame by frame with the selected scenes.

In prestripping time code onto the videotape, certain requirements must be met or extreme difficulties may arise within the postproduction phase. It is important that each roll of videotape be stripped with uninterrupted, ascending time code addresses. If the numbers do not follow the ascending sequence or are not continuous, the computer may not be able to decide which direction to go upon receiving a manual or automatic "go to" command.

At no point on any video or audio recording tape should the time code be allowed to pass through the address 00:00:00:00, when any form of synchronous or automated control is to be employed. The following example illustrates how an edit controller would handle such an occurrence. Suppose the following is true for a tape:

Time code begin = 23:46:57:19
Time code end = 00:15:38:06
Tape location = 00:03:48:10

If the edit controller is commanded to "go to" 23:56:14:00, one would expect it to respond by placing the transport into the rewind/search mode. However, since the tape location is presently at an address greater than 00:00:00:00, the controller will instead respond by placing the transport into the fast forward mode and continuing in this mode until the tape is wound off the supply reel of the transport.

A general convention is to begin the time code on an edited master at 00:58:00:00. This provides for a two-minute preroll time (with the program intro occurring at 01:00:00:00), ample time for video bars, audio tones, and slates, and for the system to achieve synchronous lock. This convention also allows for changes to be easily made to the program intro without complications.

Off-Line Video Editing

The main product of an off-line video editing session is an edit decision list and possibly a sample cut videotape.

In off-line editing, 3/4-inch or 1/2-inch videocassette dupes are made from the original field production videotapes. This format best allows for locating video and/or audio edits in order to produce a comprehensive EDL. Time code may be recorded onto these videocassettes in two ways. A time code character inserter may be connected into the recording chain, allowing the VCR to record a visible readout or "burned-in" display of the time code over the recorded video image (in addition to the recorded LTC). Time code may instead be recorded onto a new videocassette in the form of VITC time code. Then, with the use of either a VITC time code reader or the burned-in time code readout, it becomes possible to identify each video frame in any visible picture mode, including still frame.

The advantage of off-line editing is the savings in costs incurred. An off-line editing room, equipped with 3/4-inch or 1/2-inch videocassette machines, may cost a third as much to set up as an on-line editing room containing 1-inch VTRs, an EDL-based controller, and effects devices. Since an off-line facility is also less expensive to operate, the video editor can afford to spend more time making creative decisions. In addition, the amount of time that has to be spent using the more costly on-line system can be greatly reduced when the edit decision list is fully calculated within the off-line facility. During the actual on-line edit session, the edit decision list can be loaded into the controller's memory, and the final version of the program will be constructed automatically or semiautomatically, with all the edit cuts, dissolves, wipes, keys, and audio changes called for within the EDL.

Videotape Editing with VITC

SMPTE/EBU longitudinal time code (LTC) has been in use in video postproduction for a number of years. Unfortunately, LTC has certain inherent characteristics that limit the performance of more recently developed video equipment such as helical scan VTRs and video edit controllers. One significant limitation is the inability of LTC to be read at slow transport speeds and in still frame. Because LTC cannot be read in these modes, it is difficult or impossible for video edit controllers to store selected edit points accurately. Frame-accurate and correctly color-framed edits therefore become time-consuming to perform. Another significant limitation is the need to use an audio track on the videotape for recording the LTC. The necessity of dedicating an audio track to LTC precludes using that track for other material, such as stereo or a translation into another language.

SMPTE and EBU jointly standardized the vertical interval time code (VITC), which can be used to overcome the limitations of LTC. VITC is recorded within the video signal in the vertical interval, where it becomes an invisible part of the picture. Along with the visible part of the picture, the vertical interval is reproduced by a helical scan VTR, even when the videotape is stopped or moving slowly. Therefore, the video edit controller is able to read VITC and locate exact edit points by frame—every time.

When VITC was first introduced, it could only be read at play speed or slower. Therefore LTC still had to be used at faster tape speeds. Both codes were therefore required for editing: VITC when making the decisions and LTC when making the edits. Recently, the ability to read VITC at high VTR speeds and to translate it to LTC at all speeds has been developed by Adams-Smith, making it no longer necessary to utilize LTC in most videotape editing.

Since VITC may only be recorded while a video signal is being recorded, it must be laid onto the source tapes during the original video shoot or by making dupes of the source tapes prior to editing. Duping to a second generation is normally not a cause of problems when it consists of the creation of 3/4-inch videocassettes for off-line editing. However, if on-line editing of 1-inch tapes is to be done, this type of duping may affect final quality. If the primary reason for using VITC during on-line editing is to free up an additional audio track, source tapes containing only LTC may be made and an edited master videotape containing only VITC may be assembled.

In the production of an edited master with VITC, the initially blank tape must be prestripped with VITC in order to utilize the techniques of insert editing. Prestripping is accomplished by recording a composite video color or black signal with VITC added. Alternatively, assemble edits may be performed with new VITC being added and recorded as each edit is carried out. In both instances, jam-sync time code must be generated. When assemble edits are carried out, new sequential VITC is recorded as each new scene is recorded. With insert edits, the jam-sync technique must be used to rerecord the VITC being erased by the edit process.

With either LTC or VITC in videotape editing, it is of extreme importance that the source VTR(s), the record VTR(s), the TBCs, the LTC or VITC generators, and the edit controller be locked to the same video sync pulse generator.

Digital Audio Editing

Audio tape may be edited by different methods depending on whether the recorded signal is analog or digital in nature.

The standard method of editing analog audio tape is to make an appropriate razor-blade splice at a 45° angle with the tape path. This allows the signals on either side of the splice to magnetically average as it passes the head gap. This technique has been used successfully for many years (Fig. 6-17A).

Fig. 6-17. Editing of audio tape.
(Courtesy of Sony Corporation of America, Inc.)

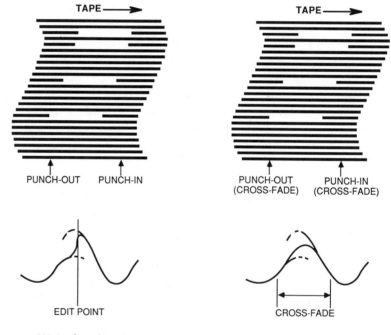

(A) Analog edit point.   (B) Digital cross-fade.

As might be expected, the editing of digitally recorded audio tape is accomplished differently than that of its analog counterpart. Digital editing employs an electronic *cross-fade* (Fig. 6-17B), which may vary in length for different formats in the manual mode from 2.5 milliseconds (PRO-DIGI) to 10 milliseconds (DASH). This computer-generated cross-fade interpolates an average level between the involved digital signals at the cut-in point, creating a continuous splice completely free of noise and distortion.

In order to take full advantage of the control of location and cross-fade within electronic editing, a choice of methods is available to the user. These are the manual digital edit and the automatic digital edit.

The Manual Digital Edit

The simplest way of performing a digital audio edit is to use the *manual digital edit*. This method, which is available only for most professional fixed-head digital ATRs, allows the operator to perform a razor-blade splice in much the same way as for an analog audio tape. In fact, this manual procedure is identical to analog editing except for two important additions: the butt splice and a safeguard against physical tape contamination.

With the manual digital edit, the splice must be oriented at a precise 90° angle to the tape length and the pieces must be joined at the edit point with a butt splice, often accomplished using a precision splice jig. The 90° angle is preferred over the standard analog 45° angle due to the extensive error detection that is necessary across a wide portion of the digital audio tape. In the reproduce mode, error detection is employed to correct for tape and system irregularities. However, as the butt splice passes the digital magnetic heads, a large number of signal errors are detected and then placed into storage in buffer memory. At this point, the error-detection system may exercise two options. When its circuitry detects an interruption of the PCM sync signal (such as when encountering a vertical butt splice), an electronic cross-fade is applied while the data are still contained within buffer memory. Should the signal error be greater than that which the error-detection circuitry can effectively handle (as would occur at any splice angle other than 90°), the audio signal output would be temporarily muted until valid data are again encountered. This is of course an undesirable condition, but is used as a final backup protection against unpredictable signal generation.

Since the density of the information recorded on digital audio tape is extremely high, the contamination of this data by dirt and oil from the hands tends to cause the error-correction circuitry to interpolate data more often than it duplicates it. It is therefore recommended that physical handling of tape be kept to a minimum and that the hands be kept very clean and free of oil during this editing procedure. Most manufacturers recommend the use of white cotton gloves and precision splice jigs to ensure that the recorded data will be kept in its most pristine form.

It is important to note that, since digitally recorded audio will not reproduce at any transport speed other than play speed, the location of digitally recorded edit points must employ the use of coincident analog cue tracks. Once located, these edit points should be marked with a new fluorescent marker (grease pencils tend to cause undue tape contamination and head wear).

The Automatic Digital Edit

When the digitized PCM information has been recorded on a rotating-head VCR or VTR (serving as a digital ATR storage medium) or when additional flexibility is required of the digital audio edit, it is necessary to utilize the *automatic digital edit*. This technique offers the advantage of allowing for variable control over cross-fade times (Fig. 6-18). The chosen cross-fade times may be tailored to best suit the audio program content, but usually range in duration from 5 milliseconds to 100 milliseconds.

Fig. 6-18. Variable cross-fade times for the automatic electronic edit. *(Courtesy of Sony Corporation of America, Inc.)*

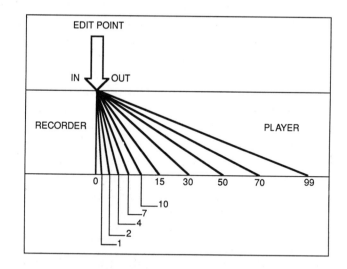

Automatic gain control may be found on most digital audio editors. This feature provides a gain offset fader, which is able to match the levels of the program material being edited together. This matching is accomplished digitally, without the operator's having to make an additional audio dub, a convenience not found with either the analog or digital manual edit. Additionally, one of the greatest benefits of this

Fig. 6-19. Diagram of the Sony DAE automatic electronic digital edit system.
(Courtesy of Sony Corporation of America, Inc.)

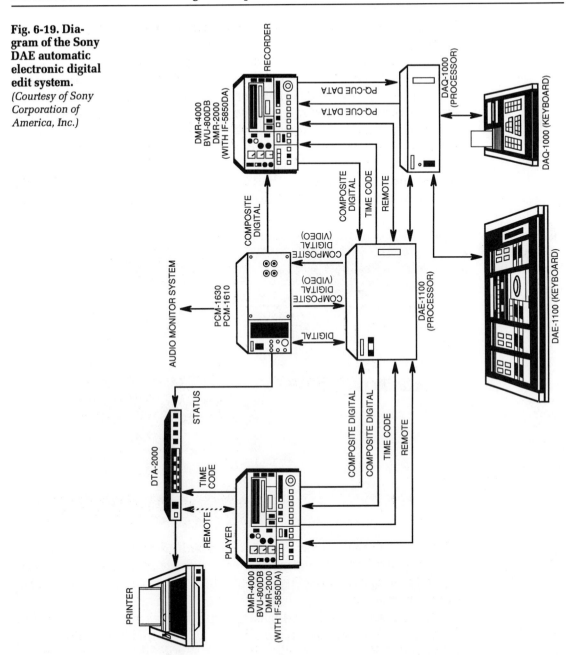

type of system is the ability to preview an audio edit in advance of actually performing it.

In theory, automatic electronic editing of audio is quite similar to electronic editing of video. However, in practice, some differences quickly become evident. The majority of present-day digital audio editors perform the automatic edit as a simple assemble edit, which is somewhat easier to perform than is its video counterpart. The automatic digital audio edit system (Fig. 6-19) employs two digital ATRs (a record and a source machine), their associated audio processors, and a digital audio edit console. The JVC AE-900V (Fig. 6-20) digital audio system, which is primarily designed to operate in conjunction with ro-

Fig. 6-20. JVC AE-900V digital audio system. *(Courtesy of JVC Company of America.)*

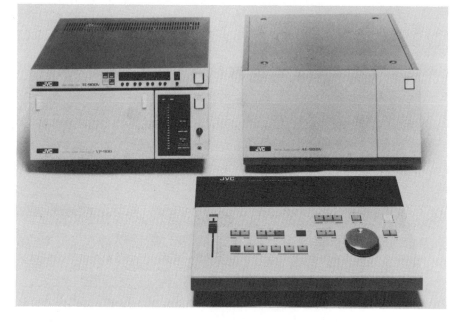

tating-head digital ATRs, operates by dividing each video head scan (with the head wheel rotating at a standard rate of 30 fr/sec) into 1470 digitally sampled words. This yields the currently accepted 44.1-kHz sampling rate.

The electronic video edit using time code is accurate to 1 video frame or 1/30 second. For the medium of video, this will yield acceptable results. For digital audio, this accuracy would result in a totally unacceptable gap and the generation of errors that would cause system shut-down at the output. For digital audio, the switching accuracy must be accurate to within 180 milliseconds. It is therefore the job of the digital audio edit controller to maintain tight control over both the record and source ATRs and to initiate the required switching functions between these machines. The digital controller is also able to maintain proprietary control over switching and ATR functions by monitoring the digital sync pulses and SMPTE time code (or a derivative of this time code).

The JVC digital audio system, which can perform both assemble and insert edits, maintains control over the ATRs through the use of a time code derivative that is printed onto line 1 of the video signal. By passing it through an external device, this derivative may be phaselocked with true time code, allowing for synchronous playback. An RS-232 serial output port enables the system to be placed under the control of a video edit controller.

A Prevention and Troubleshooting of Time Code Problems

This appendix outlines many of the potential problems that are regularly encountered in the use of time code (TC) for both audio and video production applications.

  I. Time code recording and reproduction problems

  A. Crosstalk problems

   1. The TC signal may have been recorded at too high a level, resulting in broadband crosstalk to adjacent ATR signal tracks. The jam-sync function may be used to lay a new TC signal over the old at a lower level (−5 VU to −10 VU).

  B. Loss of system code or system displaying invalid IC data

   1. The TC signal may have been recorded at too low a level to trigger the synchronizer/reader. The jam-sync function should be used to restrip TC at a higher level.

   2. The synchronizer/reader may be unable to decode longitudinal time code (LTC) at shuttle speeds approaching 1/20 of play speed or lower. Increase transport shuttle speed. For productions employing videotape, restrip using vertical interval time code (VITC) where applicable, allowing for TC monitoring in slow or still-frame reproduction modes.

C. TC distortions introduced into the record/reproduce chain

    1. The TC signal may have been recorded at too high a level, resulting in erratic and nonsynchronous operation. The jam-sync function should be used to restrip TC at a lower level.

    2. Due to the nonlinearities of the magnetic recording process, the binary-encoded TC signal may deteriorate after having been dubbed over several generations. It is therefore wise to employ momentary jam-sync for the generation of fresh code.

D. Loss of system code at high shuttle speeds

    1. In order for a TC reader to decode at high shuttle speeds, it is necessary for the designated channel to contain a sufficient bandwidth to reproduce at transport speeds of up to 100 times play speed. The TC track's reproduce amplifier must be modified in order to provide for a wider frequency bandwidth.

    2. Modern LTC readers allow tape position to be monitored at high shuttle speeds, with the tape head lifters activated through the use of a control track pulse. However, on machines that are able to read true TC in the fast shuttle mode, it is necessary that the tape lifters be disengaged. The tape transport must be modified in order to defeat the tape lifters while in the fast wind or search mode.

II. Jam-sync performance problems

A. Jam-synced material will not sync to VTR

    1. Jam-synced output is not locked to a master video sync reference. For example, if the TC generator feeding an ATR is referenced to line (60 Hz) and the VTR sync source is that of the house sync generator (59.94 Hz), a nonsynchronous condition will result, creating timing and/or frame offset errors such as out-of-phase TC where two VTRs are locked to the wrong color frame. All TC generators that are to be used in video or audio-for-video production *must* be locked to the house video sync generator in order to ensure smooth and synchronous time code.

B. Jam-synced TC output signal offset from original TC value

1. TC input is nonsynchronous (not locked to house sync), resulting in loss of editing ability with frame accuracy. Rejam source material with reference to house sync.

2. Momentary jam-sync function will not properly track discontinuous TC, resulting in inaccurate frame counts. Rejam using the continuous jam-sync mode.

III. TC problems encountered in editing

A. Controller cues tape in the wrong direction

1. TC crosses midnight (00:00:00:00). This condition is to be avoided since most synchronizers or controllers cannot discern direction once midnight is crossed. Set TC generator to an appropriate time clear of midnight.

2. When TC is not ascending (if TC numbers are repeated several times on a reel), this may cause confusion or problems in production. One solution is to manually cue the tape. Another is to restrip the tape with ascending TC. (Note: In order to maintain a correct EDL-to-frame reference in postproduction, the reel must be stripped with new time code *before* the EDL has been assembled within the spotting phase.)

B. Frame rate standardization

1. Next to level problems, the most common problem with TC in production is frame rate standardization. It is wise to consult with postproduction or broadcast clients beforehand, so that material received will be in keeping with in-house standards.

2. The vast majority of postproduction facilities use the nondrop mode (29.97 fr/s) as a standard, so as not to encounter missing frame numbers within the edit phase. Broadcasters, however, often standardize on the drop-frame mode, allowing for program running time to match the actual clock-on-the-wall time.

C. Program begins at midnight

1. It is difficult, if not impossible, to implement edits that begin at or around midnight. For this reason, a

convention of beginning an edited master at 00:58:00:00 has become established to some degree. This convention provides for 2 minutes of silence and reference signals prior to the start of the program, for example:

00:58:00:00 = black and silence
00:58:30:00 = bars/tone in
00:59:30:00 = bars/tone out
00:59:40:00 = slate
00:59:50:00 = 10-second countdown
01:00:00:00 = fade from black

B Definition and Standards of SMPTE, VITC, and Pilot System

Section I: SMPTE 12M (Revision and Redesignation of ANSI V98.12M-1981); Proposed American National Standard for Television—Time and Control Code—Video and Audio Tape for 525-Line/60-Field Systems

1. Scope.

   1.1. The first part of this standard specifies a format and modulation method for a digital code to be recorded on a longitudinal track of video and audio magnetic tape recorders. The code is to be used for timing and control purposes.

   1.2. The second part specifies the digital format to be inserted into the television signal vertical interval to be used for timing and control purposes in video magnetic tape recorders. This part also specifies the location of the code within the television baseband signal and its relationship to other components of the television signal and to the longitudinal track code described in the first part of this standard.

2. Reference standards. The following standards are intended to be used in conjunction with this standard:

2.1. EIA Industrial Electronics Tentative Standard No. 1, Color Television Studio Picture Line Amplifier Output Drawing.

2.2. International Standard ISO 646-1983, Information Processing—ISO 7-bit Coded Character Set for Information Interchange.

2.3. International Standard ISO 2022-1982, Information Processing—ISO 7-bit and 8-bit Coded Character Sets—Code Extension Techniques.

3. Longitudinal Track Applications.

3.1. Modulation method. The modulation method shall be such that a transition occurs at the beginning of every bit period. "One" is represented by a second transition one half of a bit period from the start of the bit. "Zero" is represented when there is no transition within the bit period (see Fig. B-1).

**Fig. B-1. Longi-
tudinal recorder
waveform.**

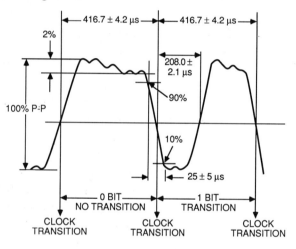

3.2. Code format.

3.2.1. Frame makeup. Each television frame shall be identified by a unique and complete address. A frame consists of two television fields or 525 horizontal lines. The frames shall be numbered successively 0 through 29, except as noted in Section 5.2.2 (drop-frame mode). If color frame identification in the code is required, the even units of frame numbers shall identify frame A and the odd units of frame numbers shall identify frame B, as defined by EIA Tentative Standard No. 1.

3.2.2. Frame address. Each address shall consist of 80 bits numbered 0 through 79.

> 3.2.2.1. Boundaries of address. The address shall start at the clock edge before the first address bit (bit 0). The bits shall be evenly spaced throughout the address period and shall occupy fully the address period, which is one frame. Consequently, the bit rate shall be 80 times the frame rate in frames per second. (See Section 3.2.1 for the definition of a television frame.)

> 3.2.2.2. Start of address. The start of the address shall occur at the beginning of line 5 in fields I and III, as defined in EIA Tentative Standard No. 1. The tolerance shall be +1 line.

3.3. Longitudinal recorder input waveform characteristics (see Fig. B-1).

3.3.1. Rise time. The rise and fall times of the clock and "one" transitions of the code pulse train shall be 25 ± 5 microseconds, measured between the 10% and 90% amplitude points on the waveform.

3.3.2. Amplitude distortion. Amplitude distortion, such as overshoot, undershoot, and tilt, shall be limited to 2% of the peak-to-peak amplitude of the code waveform.

3.3.3. Time of transitions. The time between clock transitions shall not vary more than 1% of the average clock period measured over at least one frame. The "one" transition shall occur halfway between the two clock transitions within 0.5% of one clock period. Measurements of these timings shall be made at half-amplitude points on the waveform.

3.4. Use of binary groups. The binary groups are intended for storage of data by the users, and the 32 bits within the 8 groups may be assigned in any manner without restriction if the character set used for the data insertion is not specified and the binary group flag bits 43 and 59 are both zero. If an 8-bit character set is used, the binary group flag bits 43 and 59 shall be set according to the following truth table:

	bit 43	**bit 59**
Character set not specified	0	0
8-bit character set	1	0
Unassigned	0	1
Unassigned	1	1

Unassigned states of the truth table cannot be used and their assignment is reserved to the SMPTE.

3.4.1. If an 8-bit character set conforming to ISO 646-1983 and ISO 2022-1982 is signaled by the binary group flag bits 43 and 59, the characters should be inserted in accordance with Fig. B-2. Information carried by the user bits is not specified.

Fig. B-2. Use of binary groups to describe ISO characters coded with 7 or 8 bits.

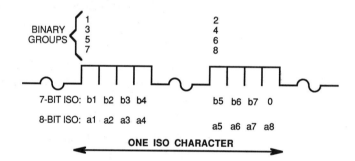

3.5. Assigned and unassigned address bits. Six bits are reserved within the address groups, 4 for identifying operational modes, 1 for biphase correction, and 1 unassigned bit reserved for future assignment and defined as zero until further specified by the SMPTE.

Bit 10—drop frame flag. If certain numbers are being dropped to resolve the difference between real time and color time, as defined in section 5.2.2, a "1" shall be recorded.

Bit 11—color frame flag. If color frame identification has been intentionally applied, as defined in section 3.2.1, a "1" shall be recorded.

Bit 27—"biphase mark" phase correction. This bit shall be put in such a status that every 80-bit word will contain an even number of logical zeros. This requirement results in the following truth table for bit 27, with the number of logical zeros in bits 0 to 63 (27 excluded) being:

Number of Zeros	bit 27
odd	1
even	0

Bits 43 and 59—binary group flag bits. These two bits shall be set in accordance with the truth table as specified in section 3.4.

Bit 58—unassigned address. "0" until assigned by the SMPTE.

The bits shall be assigned as shown in Fig. B-3 and described below:

0-3	Units of frames
4-7	First binary group
8-9	Tens of frames
10	Drop frame flag (see section 3.7)
11	Color frame flag (see section 3.7)
12-15	Second binary group
16-19	Units of seconds
20-23	Third binary group
24-26	Tens of seconds
27	Biphase mark phase correction bit (see section 3.5)
28-31	Fourth binary group
32-35	Units of minutes
36-39	Fifth binary group
40-42	Tens of minutes
43	Binary group flag bit (see section 3.4)
44-47	Sixth binary group
48-51	Units of hours
52-55	Seventh binary group
56-57	Tens of hours
58	Unassigned address bit (0 until assigned by the SMPTE)
59	Binary group flag bit (see section 3.4)
60-63	Eighth binary group
64-79	Synchronizing word
64-65	Fixed zero
66-77	Fixed one
78	Fixed zero
79	Fixed one

Fig. B-3. Longitudinal bit assignment.

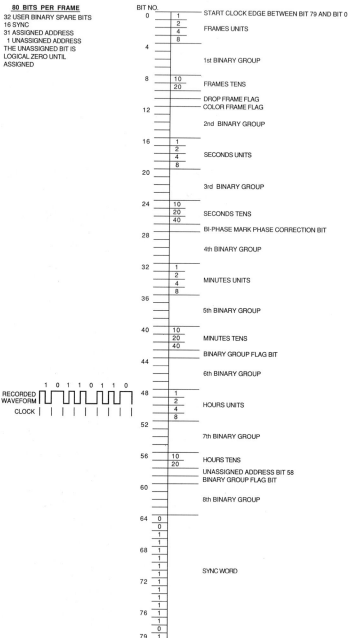

4. Vertical interval application.

    4.1. Modulation method. The modulation method shall be such that each state of the signal corresponds to a binary state and a transition occurs only when there is a change in the data contained in adjacent bit cells from "1" to "0" or "0" to "1." No transitions shall occur when adjacent bits contain the same data. Synchronization bit pairs shall be inserted as required in section 4.2.3 (modified NRZ).

4.2. Format.

   4.2.1. Makeup. The frames shall be numbered successively 0 through 29, except as noted in section 5.2.2 (drop-frame), with field identification as specified in section 4.4.

   4.2.1.1. The address recorded in each field shall relate directly to the field/frame identification as set forth in EIA Tentative Standard No. 1, and shall be related to the longitudinal code as shown in Fig. B-4.

   4.2.1.2. Bit rate. The bit rate, $F(e)$, at which the address is generated shall be as follows:

$$F(e) = F(h) \times \frac{455}{4} + 200 \text{ Hz}$$

where

   $F(h)$ is the horizontal line rate.

   4.2.1.3. Recorder input waveform characteristics. The baseband video signal after address insertion shall be specified as shown in Fig. B-5.

   4.2.2. Address. Each address shall consist of 90 bits numbered 0 through 89.

   4.2.2.1. Boundaries of address. The address shall start at the leading edge of the first synchronizing bit (bit 0). The bits shall be evenly spaced throughout the address period and shall occupy fully the address period, which is 50.286 milliseconds nominal in duration.

   4.2.2.2. Timing of the start of address. The half-amplitude point of bit 0 shall occur not earlier than 10.0 milliseconds following the half-amplitude point of the leading edge of the line synchronizing pulse. The half-amplitude point of the trailing edge of bit 89 (logical 1) shall occur not later than 2.1 milliseconds before the half-

Fig. B-4. Relationship of VITC to LTC.

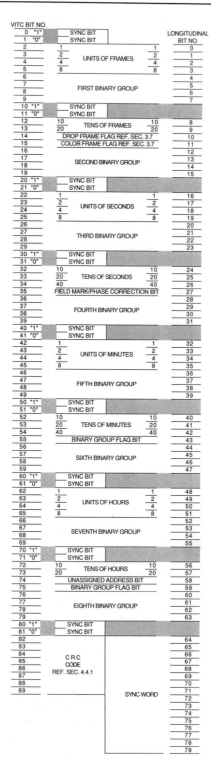

Fig. B-5. Vertical interval recorder waveform.

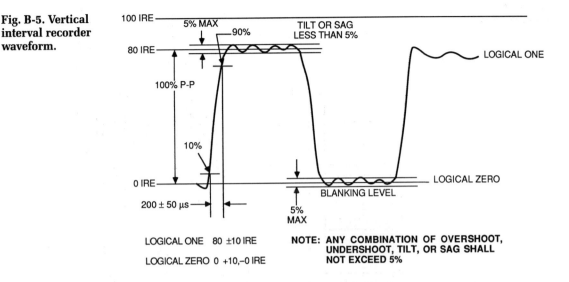

LOGICAL ONE   80 ±10 IRE

LOGICAL ZERO   0 +10,–0 IRE

NOTE: **ANY COMBINATION OF OVERSHOOT, UNDERSHOOT, TILT, OR SAG SHALL NOT EXCEED 5%**

amplitude point of the leading edge of the following line synchronizing pulse (see Fig. B-6).

4.2.2.3. Location of the address code signal in the vertical interval. The address code signal, generated at the bit rate, $F(e)$, shall be inserted on two nonadjacent lines of the vertical interval in both fields. Insertion of the address code shall not be earlier than line 10 or later than line 20, as defined in EIA Tentative Standard No. 1. The address code shall be on the same lines in all fields for a given recording. User bits shall be the same in both fields of a frame to avoid confusion when transferring from the vertical interval to the longitudinal code.

4.2.3. The bits shall be assigned as shown in Fig. B-6.

4.3. Use of the binary groups. The binary groups are intended for storage of data by the users, and the 32 bits within the 8 groups may be assigned in any manner without restriction if the character set used for the data insertion is not specified and the binary group flag bits 55 and 75 are both zero. If an 8-bit character set is used, the binary group flag bits 55 and 75 shall be set according to the following truth table:

**Fig. B-6. Address
bit assignment.**

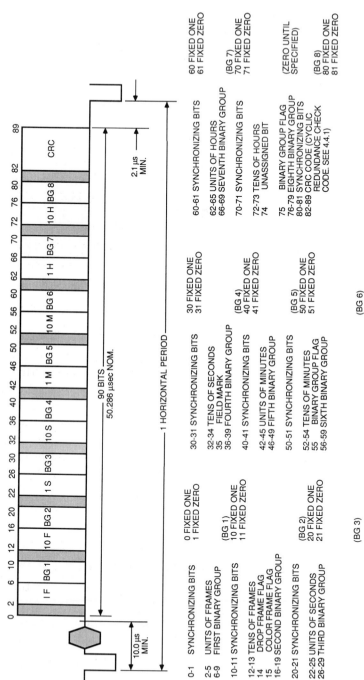

	bit 55	bit 75
Character set not specified	0	0
8-bit character set	1	0
Unassigned	0	1
Unassigned	1	1

Unassigned states of the truth table cannot be used and their assignment is reserved to the SMPTE.

4.3.1. If an 8-bit character set conforming to ISO 646-1983 and ISO 2022-1982 is signaled by the binary group flag bits 55 and 75, the characters should be inserted in accordance with Fig. B-2. Information carried by the user bits is not subject to any regulation.

4.4. Assigned and unassigned address bits. Six bits are reserved within the address groups, four for identifying operational modes, one for biphase correction, and one unassigned bit reserved for future assignment and defined as zero until further specified by the SMPTE.

Bit 14—drop frame flag. If certain numbers are being dropped to resolve the difference between real time and color time, as defined in section 5.2.2, a "1" shall be recorded.

Bit 15—color frame flag. If color frame identification has been applied intentionally, a "1" shall be recorded. Color frame identification of the code is defined as the even units of frame numbers identifying frame A and the odd units of frame numbers identifying frame B. Frames A and B correspond to color frames A and B as defined by EIA Tentative Standard No. 1.

Bit 35—field mark. Field identification shall be recorded as follows: A "0" shall represent the field in which the first preequalizing pulse follows the preceding horizontal sync pulse by a whole line. This corresponds to monochrome field I and color field I or III. A "1" shall represent the field in which the first preequalizing pulse follows the preceding horizontal sync pulse by a half line. This corresponds to monochrome field II and color field II or IV. Color fields I and III and II and IV are defined in EIA Tentative Standard No. 1.

Bits 55 and 75—binary group flag bits. These two bits shall be set in accordance with the truth table specified in section 4.3.

Bit 74—unassigned address. "0" until assigned by the SMPTE.

4.4.1. Cyclic redundance check code. Eight bits, 82 to 89, are set aside at the end of the code to provide for error detection through a check for cyclic redundance. The generating polynomial of the cyclic redundance check, $G(X)$, will be applied to all bits from 0 to 81 inclusive and shall be as follows:

$$G(X) = X^8 + 1$$

The received data divided by the generating polynomial shall result in a remainder of all zeros when no error exists in the received data.

5. Time discrepancies.

    5.1. Definitions of real and color time.

        5.1.1. One-second real time is defined as the time elapsed during the scanning of 60 fields (or any multiple thereof) in an ideal television system at a vertical field rate of exactly 60 fields per second.

        5.1.2. One-second color time is defined as the time elapsed during the scanning of 60 fields (or any multiple thereof) in a color television system at a vertical field rate of approximately 59.94 fields per second.

    5.2. Because the vertical field rate of a color signal is approximately 59.94 fields per second, straightforward counting at 30 frames per second (60 fields per second) will yield an error of +108 frames (+216 fields), approximately equivalent to a timing error of +3.6 seconds in 1 hour of running time. For correction of this discrepancy, two methods of operation are allowed:

        5.2.1. Drop-frame or compensated mode. To resolve the color time error, the first two frame numbers (0 and 1) at the start of each minute, except minutes 0, 10, 20, 30, 40, and 50, shall be omitted from the count. When this mode is used, the drop frame flag of each address shall be a "1" as specified in sections 3.5 and 4.4.

6. Structure of the address bit.

>   6.1. The basic structure of the address is based upon the binary coded decimal (BCD) system. Because the count in some cases does not rise to 9, conservation of bits is achieved because 4 bits are not needed as in an ordinary BCD code. (The 24-hour clock system is used; 2:00 p.m. is 1400 hours.)
>
>   6.2. Longitudinal track and vertical interval applications. Assignment of bits and binary coded decimal arrangements for both applications are shown in Table B-1.

Table B-1. Longitudinal Track and Vertical Interval Structure

STRUCTURAL MEMBER	ASSIGNMENTS OF BITS		BINARY CODED DECIMAL (BCD)		
	LONGITUDINAL	VIT	NO. BITS	ARRANGEMENT	COUNT
Units frames	0 – 3	2 – 5	4	1 2 4 8	0 – 9
Tens frames	8 – 9	12 – 13	2	1 2	0 – 2
Units seconds	16 – 19	22 – 25	4	1 2 4 8	0 – 9
Tens seconds	24 – 26	32 – 34	3	1 2 4	0 – 5
Units minutes	32 – 35	42 – 45	4	1 2 4 8	0 – 9
Tens minutes	40 – 42	52 – 54	3	1 2 4	0 – 5
Units hours	48 – 51	62 – 65	4	1 2 4 8	0 – 9
Tens hours	56 – 57	72 – 73	2	1 2	0 – 2

Section II: SMPTE Recommended Practice 138; Control Message Architecture (for ESbus Control Communications)

1. General.

>   1.1. Scope. This practice defines the architecture of the control message language used within a general-purpose communications channel of an interface system (ESbus) that transports data and control signals between equipment utilized in the production, postproduction, and/or transmission of visual and aural information. It is intended that the language described in this practice be utilized when constructing messages used as part of an overall system, allowing interconnection of programmable and nonprogrammable equipment as required to configure an operational system with a defined function and to allow rapid reconfiguration of a system to provide more than one defined function utilizing a given group of equipment.

Proposed SMPTE Recommended Practice 138 is reprinted by permission of the Society of Motion Picture and Television Engineers. This committee proposal is published by the SMPTE for comment only and is subject to change.

1.1.1. Control message language is composed of vocabulary, syntax, and semantics expressed in terms of tokens, rules, and actions, respectively.

1.1.2. The primary intent of this practice is to define the architecture of the messages to be transmitted within the supervisory protocol of the communications channel for the purpose of controlling equipment by external means. Syntax is the set of rules that shall be applied to the vocabulary (tokens) to construct control messages. (The content of the vocabulary and its semantics, being specific to the type of generic equipment, are defined elsewhere.) This practice, or sections thereof, may be applied to the interconnection of elements within an item of equipment.

1.2. Definitions. For the purpose of this practice, the following definitions shall apply:

Virtual machine—A logical device consisting of a single device or communicating devices that respond in essence or effect as a generic type of equipment, for example, VTR, video switcher, or Telecine.

Virtual circuit—A transparent, logical, communications connection between virtual machines. The communications path, in reality, passes through other levels and is propagated over a physical medium.

2. Message structure.

2.1. Architecture. The message architecture described in this practice is prepared broadly on the principles of communications levels. This architecture follows a logical structure and is defined in terms of a virtual machine. Messages are of variable length according to function. Complex functions may be divided into basic functions, transmitted as a sequence of shorter messages for execution in the virtual machine.

2.2. Virtual machine. All messages pertaining to generic types of equipment shall be defined in terms of the virtual machine. Utilization of the virtual machine concept in defining messages provides a message architecture that is independent of machine-specific characteristics.

3. Control message classification.

    3.1. Control messages are classified in accordance with Fig. B-7.

Fig. B-7. Message classification.

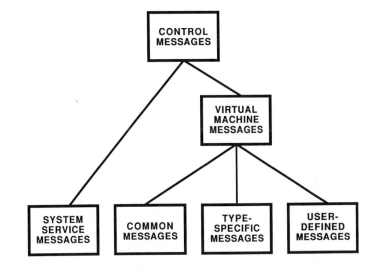

    3.1.1. Virtual machine messages are used to pass commands and responses between virtual machines. Virtual machine messages are those initiated by a controlling device with responses originating in the controlled device. Receipt of a virtual machine message shall result in a defined action and/or response by the virtual machine. Virtual machine messages may be subdivided into:

        3.1.1.1. Common messages whose coding is reserved to provide for functions of general application, for example, procedures, reference time functions, and reset.

        3.1.1.2. Type-specific messages are applicable to specific generic categories of equipment.

        3.1.1.3. User-defined messages implement special functions that are not included in the type-specific message set.

    3.1.2. System service messages are messages other than virtual machine messages.

    3.2. Virtual machine message subsets. A separate and distinct subset of virtual machine messages shall be specified for each type of virtual machine (VTR, Telecine, ATR, graphics generator, etc.). Said subset, termed a dialect, shall comprise common messages, type-specific messages, and, optionally, user-defined messages.

3.2.1. Common messages. Resident machine messages that are in all virtual machine dialects but not necessarily operative in all virtual machines, whose coding is reserved to provide for functions of general applications.

3.2.2. Type-specific messages. Virtual machine messages that are defined in virtual machine dialect recommended practices.

3.2.3. User-defined messages. Virtual machine messages that are unique to the type (manufacturer, model, version, S/N, etc.) of the specific machine. Although the definition and/or documentation of user-defined messages is considered outside the scope of this practice, the structure of such messages shall conform to the message architecture as defined herein.

4. Control message construction.

4.1. Syntax. System service and virtual machine messages are uniformly constructed with the following syntax:

$$\text{MESSAGE} = \text{KEYWORD} \, (+ \, \text{ARGUMENT})$$

where the keyword characterizes the function to be performed and the argument contains the parameters, where necessary, to perform that function. A parameter has the following syntax:

$$\text{PARAMETER} = (\text{NAME} \, +) \, \text{VALUE (S)}$$

The name may be implied with the use of specific keywords and in such cases is therefore not required. The length and format of the value (or values) is defined by the name (or implied name). No restriction is placed on the possible concatenation of parameter values.

4.2. Message formats. All control messages are formed as groups of integral bytes. The first byte of each message shall be the keyword. A keyword specification defines the format of its arguments, although no mathematical relationship is intended between the bit pattern of the keyword and the format. Messages are constructed in one of the following formats:

Format 1. Message = <Keyword>
Format 2. Message = <Keyword> <Parameter List>

where:

<Parameter List> = <Parameter>

or:

<Parameter List>=<Begin><Parameter Group><End>

where:

<Parameter Group> = <Parameter>

or:

<Parameter Group> = <Parameter Group> <Parameter>

where:

<Parameter>=<Parameter Value> ... <Parameter Value>

or:

<Parameter>=<Parameter Name>
        =<Parameter Value> ... <Parameter Value>

The appropriate message format can be selected by means of the decision tree given in Fig. B-8.

5. Message coding.

   5.1. Identical or similar functions on equipment of differing generic types should be effected by the same keyword bit pattern.

   5.2. Parameter values. Messages may contain parameters as an essential part. All parameters are classified as follows:

      5.2.1. Logical parameter values. Parameters representing any abstract function(s) that may be expressed by a simple binary state of 1 (true) or 0 (false), such as tally on/off or yes/no. The minimum code length for a single logical parameter is 1 byte. Individual logical parameters can be assembled, where applicable, into groups to form bit-specific bytes for transmission purposes.

Fig. B-8. Decision tree for selection of message format.

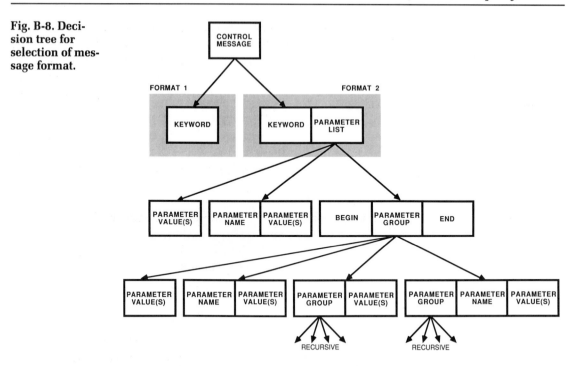

5.2.2. Numerical parameter values. Parameters representing a numeric value and consisting of the following:

Unsigned number parameters—Parameters representing any numeric value without polarity.

Signed number parameters—Parameters representing any numeric value with polarity.

Time code parameter values—Time is indicated as a 4-byte quantity. Parameters representing hours:minutes:seconds:frames are expressed in BCD in that order. The hex "40" bit of the frame's byte will be set to one (1) in drop-frame (compensated) mode. In nondrop frame (uncompensated) mode and all other time code standards, this bit will be zero (0). In all standards, the hex "80" bit of the second's byte will be set to zero (0) to indicate monochrome field 1, or color field 1, 3, 5, or 7. This bit set to one (1) will indicate monochrome field 2 or color field 2, 4, 6, or 8. Unused bits are reserved and are set to zero (0) until defined. (See Fig. B-3.)

High-resolution time code parameter values— High-resolution time is indicated as a 6-byte quantity. The first four bytes are exactly the same as time parameter values. The two remaining bytes express fractions of frames as a 16-bit unsigned number. (See Fig. B-9).

Fig. B-9. High-resolution time parameter format.

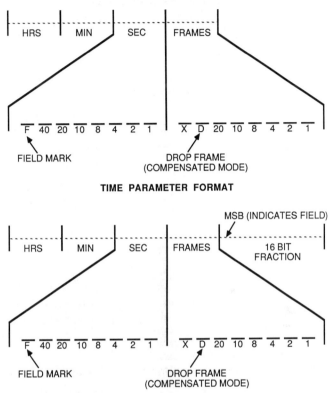

5.2.3. Literal parameters are parameters based, in general, on ASCII characters.

5.2.4. Raw data parameters are parameters based on a free-form data stream. Raw data parameters must provide for byte transparency to the lower layers. The first byte of a raw data parameter shall be a byte count.

Section III: Remote Control System (ESbus) for Broadcast Audio Tape Recorder Type-Specific Messages (SMPTE Subcommittee Draft 7.01, written by Ted Staros)

This section contains a general explanation of some of the concepts used in the formulation of the ATR type-specific message set. It constitutes tutorial information and assumes a basic understanding of the following ESbus topics (some of which are covered in Section II):

ESbus system overview
Control message architecture
Supervisory protocol
Tributary interconnection
Electrical and mechanical characteristics
System service and common messages

The ATR type-specific dialect shares many conceptual constructs with the VTR type-specific dialect; however, there are significant differences in the form and the function of their command structures. The reader is cautioned not to assume that a transparency of control messages between the dialects has been provided. System control designers who intend to exert control over both type-specific devices are urged to review these differences carefully. The dialects have been constructed to provide control in a manner most meaningful to the respective usages of the devices, while remaining true to the roots of their respective control heritages.

Some conventions for this section are as follows: Acronyms and abbreviations are written in uppercase characters, for example,

audio tape recorder = ATR
tape motion state = TMS
information field = I/F

Message keywords and names of information fields are also written in uppercase characters, for example, RECORD TRANSITION and RE-QUESTED OFFSET. These commands, keywords, and information field names are used within this section to imply requested action, information field identity, and, in turn, the information field contents of the virtual machine. To improve the readability of this material, these terms are used in the context of the presentation.

1. Command keywords and information fields. ATR-specific commands affect conditions or selection of characteristics peculiar to the virtual ATR machine. Commands that direct nonmutually exclusive conditions have corresponding information fields. When an information field is tallied, the response is in the same format as that of its respective command. Commands that direct mutually exclusive conditions may share a common information field.

2. Transport motion process and state control. The transport mechanism of an ATR is considered as a separate state machine. The commands that control transport functions are subsets within the ATR-specific message set. These are called tape motion process (TMP) and tape motion state (TMS) commands. Each TMS command causes a transition into a transport state and ceases the previous state. Tape motion process commands are overriding control commands that cause the control device to automatically choose its own tape motion state to achieve the desired result. The tape motion state will be reflected in the TMS tally, as though that TMS had been issued.

   2.1. TMP commands include TARGET, SEARCH, PRE-ROLL, SEARCH, and CHASE. All tape motion process commands are marked TMP in the index list and in the command description.

   2.2. TMS commands include STOP, PLAY, SHUTTLE, SYNC, and LOCK. All tape motion state commands are marked TMS in the index list and in the command description.

   2.3. TMP I/F tallies. This information field indicates the correct tape motion process. Since these mutually exclusive processes are commanded by the TMP commands, the code for the corresponding TMP keyword is used to identify them individually. An additional byte tallies the level of success, that is, whether the command process is still in progress or has already accomplished its task, successfully or not.

   2.4. TMS I/F tallies. This information field indicates the correct state of the transport. Since these mutually exclusive states are commanded by the TMS commands, the code of the corresponding TMS keyword is used to identify them individually. An additional byte tallies the level of success, that is, whether the command process is still in progress or has already accomplished its task, successfully or not.

3. Audio record commands (ARCs) and tallies. The recording function of the tape machine is controlled and tallied by the following keywords and I/Fs, respectively:

       REHEARSE SELECT   REHEARSE TALLY
       RECORD TRANSITION   RECORD TALLY

RECORD EXIT   —
RECORD READY SELECT   RECORD READY TALLY

   3.1. REHEARSE SELECT provides a means to identify tracks
that will, when subsequently commanded to enter a recording condition, mimic a record operation with regard
to its respective pending audio monitor commands
(AMCs).

   3.2. RECORD EXIT causes all recording tracks to exit from a
recording condition.

   3.3. RECORD READY SELECT provides a means to identify
tracks that will enter and exit a recording condition upon
the receipt of a record transition.

4. Audio monitor commands (AMCs) and tallies. The manner in
which the audio line output source selections are made may be
controlled and tallied by the following keywords and I/Fs,
respectively:

GLOBAL MONITOR SELECT   GLOBAL MONITOR TALLY
EXCLUSIVE SYNC SELECT   EXCLUSIVE SYNC TALLY
SYNC INPUT SELECT   SYNC INPUT TALLY

   4.1. GLOBAL MONITOR SELECT causes all audio channels to
present either playback, synchronous playback (sync), or
input signals to their respective line outputs.

   4.2. EXCLUSIVE SYNC SELECT provides a means of selecting
individual audio channels that will, excluding any
GLOBAL MONITOR SELECT, present synchronous playback at the line output in accordance with the SYNC
INPUT I/F.

   4.3. SYNC INPUT SELECT provides a means of choosing the
monitor switching methods used during record-related
functions. These monitor switching functions are restricted to those channels selected for synchronous
playback.

5. Velocity arguments. Some commands require a speed specification, which is carried by a command in the form of a 3-byte parameter. This parameter is intended to define the direction and
absolute value of the speed that should be achieved as closely as
possible by the ATR. This speed is referenced in terms of the
standard play speed as defined by the FIXED SPEED SELECT

I/F. All commands with a velocity parameter use the same format and coding. This is a 3-byte signed number with a scale range defined such that:

000000 hex represents a stopped condition
010000 hex represents standard forward play speed
7F0000 hex represents 127.00 times forward play speed
810000 hex represents standard reverse play speed

This allows, theoretically, for a resolution equaling 1/65,536 of standard play speed, giving an effective speed argument range of ± 127.9998 times standard speed.

6. Track selection arguments. Some commands and information fields refer to one or more tracks (or channels) of the tape machine. The format used is the same in all cases and is defined as an 8-byte map. This allows for up to 64 tracks to be controlled. The command keywords and I/Fs that utilize this track-specific mapping are:

REHEARSE SELECT	REHEARSE TALLY
RECORD READY SELECT	RECORD READY TALLY
EXCLUSIVE SYNC SELECT	EXCLUDING SYNC TALLY
—	RECORD TALLY

7. Tape code identity. There are currently two means of referencing tape location. These are INTERNAL LTC (longitudinal time code from tape) and TAPE TIMER. Separate information fields currently exist for both internal LTC and tape timing location. The content of the selected tape code I/F, however, is chosen by the TAPE CODE SELECT command.

   7.1. A TARGET SEARCH command would cause the controlled device to locate to a position on the tape referenced to the SELECTED TAPE CODE.

   7.2. The PRE-ROLL SEARCH COMMAND would always reference to the INTERNAL LTC, regardless of the SELECTED TAPE CODE.

8. Achieving and maintaining synchronization. Synchronization requires the controlled device to maintain a particular time relationship between its INTERNAL LTC and some external reference. This relationship is usually restricted to some speed range around the nominal FIXED SPEED. The external reference to which synchronization is achieved and maintained may be se-

lected from a number of alternative sources. This is accomplished with the LOCK MODE SELECT command. The LOCK command enables the process of synchronization. Although the LOCK command may be initiated to synchronize machines with no specification setup, critical performance values within the LOCK command require the following:

A specified EXTERNAL TIME CODE ("when")
A specified point on the tape ("where")
A selected external reference ("how")

8.1. "When"—This point in time is defined by the specification of the LOCK TIME I/F. This refers to a time, defined by the EXTERNAL TIME CODE, at which synchronization is assured between the EXTERNAL TIME CODE and the controlled device's INTERNAL LTC.

8.2. "Where"—This is a point on the tape called the "local lock point," which may be characterized by two independent specifications. These are the aforementioned LOCK TIME I/F and the REQUESTED OFFSET I/F.

8.2.1. The REQUESTED OFFSET I/F specifies the longitudinal time relationship between the EXTERNAL TIME CODE and the control device's INTERNAL LTC. This REQUESTED OFFSET is maintained during successful synchronous operation. (Note: A related information field, the ACTUAL OFFSET I/F is provided so that tallies of INTERNAL LTC minus the external time code may be facilitated.)

8.2.2. The local lock point may be calculated as the sum of the LOCK TIME I/F and the REQUESTED OFFSET I/F.

8.3. "How."

8.3.1. Lock modes. LOCK MODE SELECT allows a choice in the manner by which synchronization is achieved and maintained. Two different classes of synchronization may be selected: absolute and free. There are four absolute modes and two free modes available for LOCK MODE SELECT.

8.3.2. Absolute modes of lock.

8.3.2.1. Absolute standard mode—Achieve lock to EXTERNAL TIME CODE, data-dependent; maintain lock, data-dependent. External LTC is selected as the source of EXTERNAL TIME CODE.

8.3.2.2. Absolute resolve mode—Achieve lock to EXTERNAL TIME CODE, data-dependent; maintain lock, data-dependent. External LTC is selected as the source of EXTERNAL TIME CODE.

8.3.2.3. Absolute video mode—Achieve lock to EXTERNAL TIME CODE, data-dependent; maintain lock to external video reference. External LTC is selected as the source of EXTERNAL TIME CODE.

8.3.2.4. Absolute VITC mode—Achieve lock to external video with VITC, data-dependent; maintain lock to external video reference. The external video VITC signal is selected as the source of EXTERNAL TIME CODE.

8.3.3. Free modes of lock.

8.3.3.1. Free resolve mode—Achieve lock to EXTERNAL TIME CODE, data-independent; maintain lock, data-independent. External LTC is selected as the source of EXTERNAL TIME CODE.

8.3.3.2. Free video mode—Achieve lock to EXTERNAL VIDEO SIGNAL; maintain lock to external video reference. The source of EXTERNAL TIME CODE is undefined.

8.3.4. Lock operation in absolute modes. Three important functions must be established before any of the absolute modes of lock may be represented.

8.3.4.1. PRE-ROLL DURATION—Contains the time used in, or needed in advance of, achieving synchronization. The PRE-ROLL DURATION I/F specifies the exact real-time period between the start of tape movement (in response to a lock command) and the moment of encountering the specified LOCK TIME. It is assumed that EXTERNAL TIME CODE is presented to the device in a real-time manner during the pre-roll period. The PRE-ROLL DURATION I/F may not be set to a value lower than the device-dependent lower limits.

8.3.4.2. PRE-ROLL SEARCH—This TMP causes the controlled ATR to move to a tape position specified by the local lock point minus the predefined PRE-ROLL DURATION plus any device specific "acceleration allowance." This position may be described as the pre-roll search point.

8.3.4.3. Lock actuation—In all absolute modes of the lock command, the condition that causes the start of tape movement to achieve and maintain synchronization is always the receipt of EXTERNAL TIME CODE of a value equal to the predefined LOCK TIME I/F minus the predefined PRE-ROLL DURATION I/F. The time at which this occurs may be termed the "lock actuation time." The source of the external time code that triggers the lock actuation may be either LTC or VITC. This choice is specified by the LOCK MODE SELECT command. All lock commands issued in any absolute mode require predefined PRE-ROLL DURATION, REQUESTED OFF-SET, and LOCK TIME I/Fs, and must be preceded with a PRE-ROLL SEARCH command.

Section IV: BASIC Pilotone/Pilot Tape Recorder System for Nagra/Kudelski Tape Recorders

1. Basic theory of a Pilot tape recorder. For usages where (phase-locked) synchronization of the recorded signal and other equipment is needed, it is necessary to use a device called a "Pilot." The Nagra recorders are, upon request, equipped with a complete Pilot system. A block diagram of the Pilot system is shown in Fig. B-10: (A) Pilot signal recording system, (B) Pilot signal playback system, (C) magnetic tape, and (D) Pilot recording/playback head.

1.1. Pilot signal recording chain. The Pilot signal recording chain is a device allowing transcription on the tape of the Pilot signal.

Fig. B-10. The
Pilot system.
*(Courtesy of
Nagra Magnetic
Recorders, Inc.)*

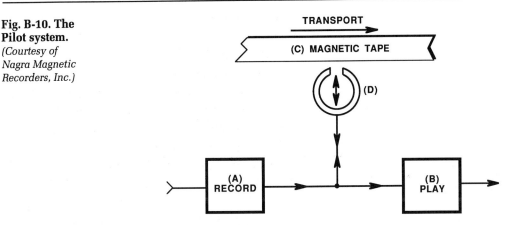

1.1.1. A diagram of the Pilot recording chain is shown in Fig. B-11: (A) VHF receiver (45 to 50 MHz), (B) Xtal Pilot oscillator (50 or 60 Hz), (C) frequency meter (50 to 60 Hz), (D) rf bias oscillator (120 kHz), (E) Pilot recording modulator amplifier, (F) Pilot recording/playback head, (G) "clapper" oscillator, (H) Xtal−Pilot-in bridging plug, (I) meter, (J) internal synchronized digital clock, (K) clock set input, (L) antenna, (M) "clap" signal input (DC pulse), (N) commentary microphone input (200 to 3500 Hz), and (O) Pilot record on/off.

Fig. B-11. The
Pilot recording
chain.
*(Courtesy of
Nagra Magnetic
Recorders, Inc.)*

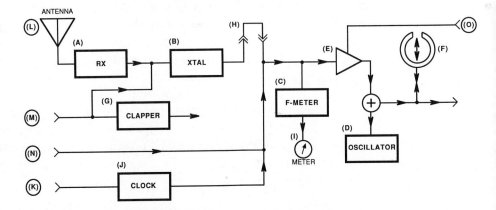

1.1.2. Working principle. The signal to be recorded on the Pilot track can be built up in several different ways:

Pilot only (50 or 60 Hz)
Pilot plus audio commentary

Audio commentary only

Real-time coding (that is, IRT or Institute fur Rundfunk Tecknik)

1.1.2.1. The 50 or 60 Hz Pilot is given by an internal quartz generator (J) that can be controlled by the "clap" and coding receiver (A) or by the "clap" signal from the input (M).

1.1.2.2. The audio commentary comes through the input (N) from a special microphone with a built-in preamplifier and dynamic compressor.

1.1.2.3. The real-time coding is generated by an integral digital quartz crystal clock, set by a master clock that can be connected to the input (K).

1.1.2.4. The "clap" can be recorded either as a short Pilot interruption or as an audio beep on the audio track according to the internal wiring; if the "clap" comes from the VHF receiver (A), it can only be recorded as a Pilot interruption.

1.1.2.5. The signal is then amplified (E) and superimposed on the bias current coming from the oscillator (D) to feed the head (F). The Pilot frequency is measured by the frequency meter (C) and displayed by the meter (I).

1.2. Pilot playback chain. The playback chain in Fig. B-12 allows the playback of the recorded Pilot signal as well as the control of the tape speed by a time reference: (A) raw Pilot output for external use, (B) motor speed correction output, (C) Pilot recording playback head, (D) playback amplifier and Pilot demodulator, (E) Pilot playback level control, (F) synchronizer (phase compensator), (G) quartz crystal oscillator, (H) Xtal−Pilot-in bridging plug, (I) frequency meter, and (J) reference frequency control.

1.2.1. Working principle.

1.2.1.1. The signal is read by the Pilot head (C), then amplified and demodulated by the stage (D).

1.2.1.2. The reference signal comes from the exterior through the Pilot-in socket or is generated internally by the oscillator (G) and fed through the plug (H).

Fig. B-12. The Pilot playback chain.
(Courtesy of Nagra Magnetic Recorders, Inc.)

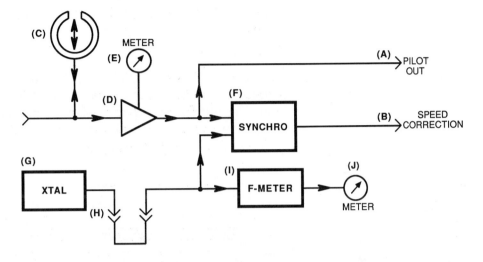

1.2.1.3. The synchronizer (F) then compares the reference frequency and the Pilot frequency read on the tape; a correction signal is compiled and fed to the motor speed control through (B).

1.2.2. Remarks. The Pilot frequencies are standardized; the most commonly used are 50 Hz for 25 fr/s (TV and cinema, Europe, IEC), 60 Hz for 24 fr/s (cinema, USA, NAB), and 60 Hz for 30 fr/s (TV, USA, NAB). However, 100, 120, 400, and 440 Hz are also found.

2. Nagra Pilotone configurations.

2.1. The Nagra 4.2.L is provided with the Neopilot synchronization system that does not allow commentary recording on real-time coding because of the superposition of the Pilot channel on the audio channel. (See Fig. B-13.)

Fig. B-13. Track configuration for Neopilot system.
(Courtesy of Nagra Magnetic Recorders, Inc.)

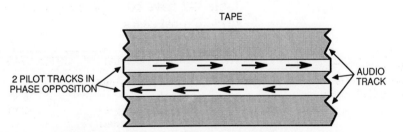

Width of the audio track = 6.30 mm
True width of the audio track = 5.30 mm (at I = 320 nWb/m)
Width of a Pilot track = 0.50 mm

Width of the two Pilot tracks = 1.00 mm (at I = 160 nWb/m)

Spacing of the Pilot tracks = 0.75 mm

2.2. The Nagra IV-S is equipped with a third-track Pilot system, which results in wider frequency and level spectra and minimal interference on the adjacent audio tracks. Digital time code signals or commentaries can easily be recorded, either in AM or FM. (See Fig. B-14.)

Fig. B-14. Third-track Pilot system tape configuration. *(Courtesy of Nagra Magnetic Recorders, Inc.)*

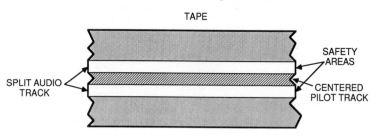

Width of the audio track = 2.00 mm

True width of the audio track = 4.00 mm (at I = 320 nWb/m)

Width of the Pilot track = 0.80 mm (at I = 320 nWb/m)

Width of the security zone = 0.75 mm

Tape width = 6.30 mm

Section V: Levels and Tape Track Configurations for Time Code

1. Time code levels.

   1.1. Currently, no industry standard level(s) exist for the recording of time code onto magnetic tape. The levels in Table B-2 have been proven over time to give the best results in most cases.

Table B-2. Optimum Recording Levels for Time Code

TAPE FORMAT	TRACK FORMAT	OPTIMUM RECORDING LEVEL
¾-inch	Audio 1 track or time code track	−5 VU to 0 VU
1-inch	Cue track or audio 3	−5 VU to −10 VU
2-inch quad	Cue track, also called auxiliary track	+3 VU to +5 VU
ATR	Edge track (highest number)	−5 VU to −10 VU

If the VTR to be used is equipped with AGC (automatic gain compensation), do not use it; override the AGC and adjust the signal gain controls manually.

2. Tape track configurations.

    2.1. The VTR and ATR tape track configurations shown in Figs. B-15 through B-19 represent the popular standard track configurations currently available for production using time code.

Fig. B-15. Track configuraton for BCN machine.

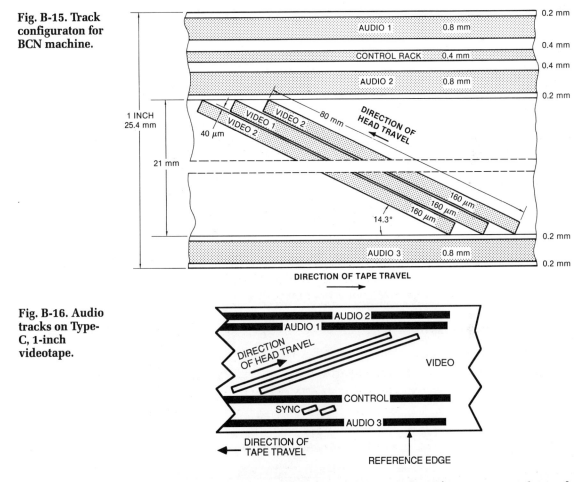

Fig. B-16. Audio tracks on Type-C, 1-inch videotape.

    2.2. Multitrack ATR time code configurations. In multitrack ATR production using time code, it is standard practice to place time code onto the highest numbered outside track available (for example, track 24 on a 24-track ATR).

        2.2.1. Often the adjacent inside track is left blank on a multitrack ATR in order to minimize crosstalk; however, proper attention to time code level adjustments will help to reduce the necessity for such a precaution.

2.3. Digital ATR time code configuration. Most professional stationary-head digital ATRs make provisions for a dedicated analog or digitally encoded time code track or memory address. Consult the machine's manual for details.

Fig. B-17. Track configuration on cassette tape.

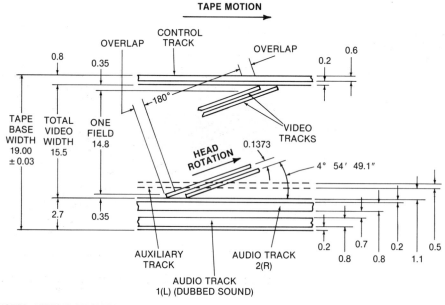

NOTES: VIEW IS OF OXIDE SIDE OF TAPE.
LINEAR DIMENSIONS IN MILLIMETERS.

Fig. B-18. Track configuration on 1/2-inch, 4-track audio tape.

Fig. B-19. Third-track Pilot system tape configuration.
(Courtesy of Nagra Magnetic Recorders, Inc.)

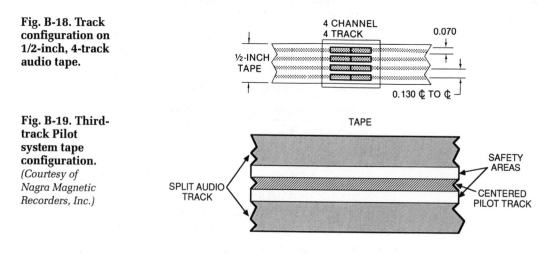

C *Glossary*

A/B roll edit Two source VTRs are operated under video and time code sync, under the direction of an edit controller. This is one production method for creating special effects from prerecorded source material by means of dissolves, wipes, etc.

Access time The time required to retrieve video information during edit or assembly.

Address lines Serial or parallel lines used exclusively for transfer or communication of address information with a memory device.

Assemble mode A A sequential assembly of the edit list. This assembly will continue in automatic mode until all edits have been assembled or until the first event with an unassigned reel number is found.

Assemble mode B In this mode, the auto-assemble process continues as long as there are reels from which to record. This mode is also known as "checkerboard" assembly. The system in many cases is more efficient in this mode.

ATR Abbreviation for audio tape recorder.

Auto-assemble Generation of an edited master by a video or audio-for-video edit controller using an existing edit decision list.

AUX, or auxiliary A channel through which a video device (video camera, film chain, etc.) may be connected to a video edit controller.

Back time Calculation of a tape in-point by finding the out-point and subtracting the duration of the edit.

BCD   Abbreviation for binary coded decimal.

Binary data   Information encoded as a series of two-state levels, usually referred to as high and low or logical 1 and logical zero.

Black burst   Provides the synchronizing signals for the system to lock onto and thus stabilize the VTRs; also known as color black, crystal sync, and edit black.

Capstan servo   An electronic circuit that controls capstan speed with enough stability so that video information can be read in correct sequence by the magnetic video heads.

Channel   A communications line.

Color frame   The video frame polarity. Color frames must alternate in polarity to keep the video signal in phase.

Control track   A synchronizing signal on the edge of the tape, which provides a reference for tracking control and tape speed.

CPU   Abbreviation for central processing unit.

Cut   The instantaneous transition from one information source to another.

Data lines   Serial or parallel lines used exclusively for transfer or communication of data in binary, ESbus, ASCII, or other encoded form.

Dissolve   An edit transition where one source of video or audio fades out while at the same time another source fades in.

Drop frame   A system for modifying the frame counting sequence to allow the time code to match the actual time of day.

Dropouts   Small bits of missing picture information, usually caused by physical imperfections in the surface of the videotape.

Dub; dupe   Terms used to describe a videotape duplicate. Dupe is the more accepted word, since dub may mean a copy in both video and audio work.

Edit   Any point on a videotape where the audio or video information has been added to, replaced, or otherwise altered from its original form.

Edit controller   Device that is central to the modern electronic video edit in that it provides both memory and control functions (it often takes the form of a microprocessor or computerized device). This central unit is able to extend limited or complete control over the peripheral devices involved within the video or audio-for-video process.

Edited master (EM)   The final edited videotape with continuous program material and time code from beginning to end.

Edit list (edit decision list, or EDL)   A record of all the edit decisions to be made in creating a video program; takes the form of a printed copy, paper tape, or floppy disk and is used to automatically assemble the program.

Edit points (edit-in, edit-out)   The beginning and ending points of a selected event within a program being assembled on magnetic tape.

Effect   A multisource transition, such as colorizing, chroma-keying, etc.

Effects switcher   An electronic switcher that includes the generation control and coordination of special effects.

Event number   A number assigned by the editor to each edit that is recorded in the EDL.

Field   One-half of a complete video scanning cycle, or 1/60 second (NTSC); one-fourth of a complete video scanning cycle, or 1/50 second (PAL/SECAM).

Field master   The raw, original program footage on videotape containing original time-of-day and shoot time code identification and, where applicable, original field audio (also known as a source master).

Frame   One complete video scanning cycle.

Generation   A number that indicates how many reproductions a dupe has gone through from the original master videotape.

Initialization   System startup, or setting of equipment switches and circuits to their beginning positions and values.

Insert edit   Type of edit in which new video material is inserted into existing material already recorded on the master tape (or into black), with the capability of returning to the existing video. No new time code or control track is recorded.

Intelligent interface   A sophisticated microprocessor-based controller of VTRs and ATRs and switchers.

Interlock   To run sound and picture together in perfect sync from separate film and/or tape transports.

Iso reels   Multiple reels of tape on which the same subject has been recorded simultaneously by different VTRs through different cameras.

Jam-sync   The process of locking a time code generator to the time code recorded on an existing tape for the purpose of recreating or extending the time code on the tape. This may be necessary because time code may become nonexistent or of poor quality after just a few generations.

Jog mode   A standard function available on broadcast-quality VTRs equipped with dynamic tracking (DT) or AST. In this mode, the VTR may be moving in fast forward, fast reverse, slow forward, or slow reverse or be in stop frame and the picture can be viewed.

Kinescope recording   Made by photographing the display of a television monitor with a motion picture camera.

Longitudinal time code (LTC)   Time code information recorded as an audio signal on a designated track of a VTR or an ATR.

Match-frame edit   An edit by which a scene already recorded on the master is continued with no apparent interruption; performed by setting the record and source in-points equal to their respective out-points for the scene that is to be extended.

Nondrop mode   A system for time code recording that retains all frame numbers in chronological order.

NTSC format   A system of coding color information for television transmission, used primarily in the United States and Japan and formulated by the National Television Standards Committee in the early 1950s.

Off-line editing   Preparation that is done to produce an edit decision list, which is later used in the auto-assemble on-line process. A videotape (sometimes called a worktape) may be produced as a by-product of off-line editing.

Offset adjustment   Accurate adjustment of the tape position of one video or audio transport relative to that of another (generally relative to the master transport). This is most often accomplished using time code and may be plus or minus a specific number of frames.

On-line editing   EDL-based computer editing that is done to produce a finished program master.

On-the-fly editing   Choosing an edit point while the tape is moving at play speed.

Open-ended edit   An edit without a defined out-point. The system will record until the record or stop button is depressed.

Preroll   The physical rolling back of both the record and source videotapes to a tape location preceding the edit cut-in point in time.

This allows for both VTRs to be synchronized for control and time code functions in advance of the actual edit.

Random access   The ability to retrieve video, audio, or data from any point on a tape, disk, or solid-state memory device.

Real time   Actual elapsed time.

Reel number   The number assigned by the operator to an audio or video reel or cassette to be used in an edit session; used for the purpose of identifying each reel or cassette on the edit list for the final assembly or for future revisions.

Servo motor   An electromechanical control device using an external reference.

Split edit   A type of edit transition where either the audio or the video of the source is delayed (not recorded) for a given period of time.

Switcher   Device used for performing simple switching (on/off) functions, transpositions, or, in complete devices, coordination of special effects.

Synchronization   The precise coincidence of two signals, pulses, or events.

Time code   SMPTE standard for encoding time for video or audio in hours:minutes:seconds:frames.

Time code generator   A signal generator designed to generate and transmit SMPTE time code.

Time code reader   A counter designed to read and display SMPTE time code.

Trim   To add to or subtract from an edit duration using a time code address entry.

Upcut   An incorrect edit in-time that requires the edit point to be trimmed back; usually the result of slow reaction times when editing on the fly.

VITC   Abbreviation for vertical interval time code; address information that is recorded within the vertical blanking segment of the video signal.

Wipe   An edit transition where one video signal replaces another video signal on the screen in some predetermined pattern.

Wipe code   A two- or three-digit numeric code used to identify a wipe pattern.

Zero frame dissolve A dissolve with a duration of zero frames, equivalent to a cut; a technique used to synchronize two source machines so that manual audio or video transitions can be made between them.

D Bibliography

Chapter 2

"Ampex Technology Report," *Video Systems*, April, 1980.

"CompuSonics DSP-2002" (technical bulletin), CompuSonics Corporation, 1985.

Modern Recording Techniques, Runstein and Huber, Howard W. Sams & Co., Inc., 1986.

The Video Guide, 3rd ed., C. Bensinger, Video Info Publications, 1982.

Video Tape Recorders, H. Kybett, Howard W. Sams and Co., Inc., 1985.

Chapter 3

"Adams-Smith System 2600" (technical bulletin), Adams-Smith, 1985.

"CASS-1" (technical bulletin), CMX/Orrox, 1985.

"EC-101 Series" (technical bulletin), Otari Corp., 1985.

"EC-400 Series" (technical bulletin), Otari Corp., 1984.

Q-Lock Manual, Audio Kinetics, Inc., 1985.

"SL 6000E" (brochure), Solid State Logic Ltd.

Time Code Handbook, Datametrics-Dresser Industries, Inc., 1982.

"TLS 4000 Synchronizer Series" (technical data), Studer/Revox of America, Inc.

Chapter 4

Modern Recording Techniques, Runstein and Huber, Howard W. Sams & Co., Inc., 1986.

"A Modified 'MS' Recording Technique for Location Recording," L. E. Weed, *Recording Engineer/Producer*, October, 1977.

"M-S Stereo: A Powerful Technique for Working in Stereo," Audio Engineering Society preprint, presented at the 69th convention, May 12–15, 1981, Los Angeles.

"Nagra AUDIO-T" (technical bulletin), Nagra Magnetic Recorders, Inc.

"Nagra IV-S" (brochure), Nagra Magnetic Recorders, Inc.

"Nagra IV-S Time Code" (brochure), Nagra Magnetic Recorders, Inc.

"Sony APR-2003" (brochure), Sony Corporation of America, 1985.

Sound Recording for Motion Pictures, A. S. Barnes & Co., Inc., 1979.

Chapter 5

Audio Sweetening for Film and TV, Hubatka, Hull, and Sanders, Tab Books, Inc., 1985.

"Buying Audio Consoles for Broadcast Video," Rupert Neve, Inc., 1986.

"Eine Kleine Sampling Music," L. Oppenheimer, *Mix* magazine, May, 1986.

"Eventide SP 2016, H 969" (brochure), Eventide, Inc.

"The Guide to Profitable In-House Production," MCI, A Division of Sony Corporation of America, 1985.

"How to Choose Equalizers," Orban Associates, 1981.

"Lexicon PCM-70, Model 2400" (brochure), Lexicon, Inc.

Making Music, Martin, Quill Books, 1983.

Modern Recording Techniques, Runstein and Huber, Howard W. Sams & Co., Inc., 1986.

"Orban 622 Parametric Equalizer" (brochure), Orban Associates, 1985.

Sound Recording for Motion Pictures, C. B. Frater, A. S. Barnes & Co., Inc., 1979.

"The Taker A/B" (brochure), Geise Electronic/ESL, Inc.

"What MIDI Means for Musicians," J. Wright, *Polyphony*, June, 1983.

Chapter 6

Audio Sweetening for Film and TV, Hubatka, Hull, and Sanders, Tab Books, Inc., 1985.

"Digital Editing—The Mitsubishi System," Mitsubishi Electric Sales America, Inc., 1986.

"Personal Notes on the Early Video Industry," personal communication from C. Ginsburg.

"Sony PCM-3324" (brochure), Sony Corporation of America, 1986.

Time Code Handbook, Datametrics-Dresser Industries, Inc., 1982.

Video Tape Recorders, H. Kybett, Howard W. Sams and Co., Inc., 1985.

Appendix B

"ANSI V98.12M-1981," American National Standards Institute, 1981.

"Basic Theory of a Pilot Tape Recorder," Nagra Magnetic Recorders, Inc.

"SMPTE Time Code Recording on Tape," (technical information), Studer International AG, 1986.

"SMPTE 12M," Society of Motion Picture and Television Engineers, 1984.

Index

A